T0225131

Series/Number 07–148

MULTIPLE TIME
SERIES MODELS

Patrick T. Brandt
University of Texas at Dallas

John T. Williams
University of California, Riverside

SAGE Publications
Thousand Oaks ■ London ■ New Delhi

For information:

Sage Publications, Inc.
2455 Teller Road
Thousand Oaks, California 91320
E-mail: order@sagepub.com

Sage Publications Ltd.
1 Oliver's Yard
55 City Road
London EC1Y 1SP
United Kingdom

Sage Publications India Pvt. Ltd.
B-42, Panchsheel Enclave
Post Box 4109
New Delhi 110 017   India

Library of Congress Cataloging-in-Publication Data

Brandt, Patrick T.
Multiple time series models / Patrick T. Brandt, John T. Williams.
    p. cm. — (Quantitative applications in the social sciences, vol. 148)
Includes bibliographical references and index.
ISBN   978-1-4129-0656-2 (pbk.)

    1. Times series analysis—Mathematical models. I. Williams, John T. II. Title. III. Series: Sage university papers series. Quantitative applications in the social sciences.
HA30.3.B73 2007
519.5'5—dc22   2006010016

This book is printed on acid-free paper.

14   15   16   17   18   10 9 8 7 6 5 4 3 2

Acquisitions Editor:	Lisa Cuevas Shaw
Associate Editor:	Margo Beth Crouppen
Editorial Assistant:	Karen Greene
Production Editor:	Melanie Birdsall
Copy Editor:	QuADS Prepress (P) Ltd.
Typesetter:	C&M Digitals (P) Ltd
Indexer:	Ellen Slavitz
Cover Designer:	Janet Foulger

CONTENTS

LIST OF FIGURES

LIST OF TABLES

SERIES EDITOR'S INTRODUCTION

Social and economic scientists have long been fascinated by and taken advantage of time series data. A first systematic exploitation of the richness of such data is William Playfair's *The Commercial and Political Atlas,* published 220 years ago and containing 43 time series graphs. By plotting the national debt of England against time, for example, Playfair could easily identify the impact of major historical events such as the accession of Queen Anne in 1701, the Spanish War of the 1730s, and the American (Revolutionary) War that began in 1775.

Playfair also used graphs with more than one time series. The graph below charts the curve of imports and that of the exports against time, clearly demonstrating that there is a relationship between the two as well as between them and time, not to mention the main purpose of Playfair's defining import-export balance against and in favor of England. Charts like this show that the two time series may not be independent processes.

Exports and Imports To and From Denmark and Norway From 1700 to 1780

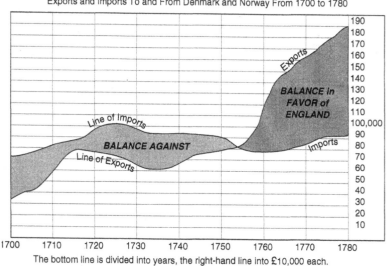

The bottom line is divided into years, the right-hand line into £10,000 each.

SOURCE: www.unc.edu/~nielsen/soci208/m2/m2033.jpg

The usefulness of time series graphs notwithstanding, Playfair's presentation leaves many questions unanswered. What caused imports or exports to rise and fall? Did the amount of imports have an impact on the amount of exports and vice versa? Perhaps most important and interesting, what changed the regime of import-export balance from against to in favor of England? Answering these questions requires a proper analysis of the dynamic simultaneous processes, and Brandt and Williams's *Multiple Time Series Models,* seen by my predecessor Michael Lewis-Beck as a worthwhile project, presents methods for such analyses.

The authors discuss the assumptions and specificities of four main approaches for time series data: autoregressive integrated moving average models, simultaneous equation systems, error correction models, and vector autoregression (VAR) models. They then focus on the details such as the specification and estimation of inference in VAR as well as tests for Granger causality and assessment of dynamic causal relationships via impulse response functions and measures of uncertainty before offering two complete examples of the VAR model of multiple time series data. A welcome addition to the series, this book complements the existing volumes of *Time Series Analysis* (No. 9), *Univariate Tests for Time Series Models* (No. 99), and *Multivariate Tests for Time Series Models* (No. 100).

—Tim Futing Liao
Series Editor

PREFACE

About This Book

This project is several years in the making. John first proposed writing this book to me in 1999 when he was teaching time series analysis and I was his teaching assistant at the Inter-University Consortium for Political and Social Research (ICPSR) Summer Program at the University of Michigan. The project sat idle until John revived the idea in 2002—although we had discussed it in the interim.

Regrettably, John passed away in September 2004 while we were working on this book. Prior to this tragic event, we had actually completed a fair amount of the outline and plan for the book. John's influence pervades this book—from the way the ideas are presented to the general outlook of how to "do" multiple time series analysis. I have tried to stay as close as possible to the original conception of the project that John and I devised.

I have been aided by a series of notes that were found in John's files. In 1994 and 1995, John did a series of lectures on vector autoregression and multiple time series models, at, among other places, the ICPSR Summer Program. Ken Bickers, Shaun Bowler, Mike McGinnis, and Matthew Potoski thankfully were able to recover these notes from his files. Regrettably, though, they had never been updated, so they only reflected the "state of the art" circa 1994. In addition, several important topics were missing from these notes, including a discussion of error correction models (Sections 1.3 and 2.7), new results on Granger causality testing (Section 2.5.4), and probability assessment methods for impulse responses (Section 2.5.8). In addition, the notes included only bullet points for the discussion in Chapter 1, not the full discussion included herein.

John and I have incurred a number of debts in the course of writing this book. We would first like to thank our friend, coauthor, and collaborator John Freeman for his steadfast role as a supporter and commentator for both of us over the years. We owe a debt of gratitude to our original Sage series editor, Michael Lewis-Beck, who worked with us to define the project and get it accepted by Sage. The current series editor, Tim Liao, has been especially helpful as I have worked to complete the book. I would like to thank the University of North Texas and its Political Science Department, which were my home when I started this project. Ken Bickers, Mike McGinnis, and Matthew Potoski also deserve special thanks for sorting out John's files and finding his notes on these topics. Without these files, completing the manuscript would have taken much longer. I am also indebted to the many students to whom John taught this material over the years at Indiana and the ICPSR Summer Program at the University of Michigan. I wish I could list all of you here, because I know how much John meant to you and you to him. Thanks are also due to Hank Heitoweit for inviting John to teach some of this material at the ICPSR Summer Program over the years and inviting me to be a teaching assistant for John's time series course in 1999. I also thank Harold Clarke, who provided a wonderful two-day forum where I could try out this material in front of a great audience of his (and my now current) colleagues and students in the fall of 2004. Mike Colaresi (one of John's former students and a friend) discussed some of John's general approach to time series with me, reminded me of some key ideas, and provided a critical reading of the manuscript. Finally, Justin Appleby, Mehmet Gurses, Thomas Sattler, Patrick Sellers, and Chu-Ping Vijverberg provided helpful feedback on the nearly completed manuscript.

Background Material and Plan of the Book

This book covers advanced topics in the analysis of multiple time series. By multiple time series, we mean time-ordered data where there is more than one (endogenous) variable of interest. For example, if one were to model the determinants of the percentage of Americans who claim to be Democrats (known as aggregate partisanship), an important explanatory factor might be the general public support for the policies of the U.S. government. But support for these policies will be colored by the same aggregate partisanship of the general public. Some of the relationships among these variables can be described by recent history (past values), whereas other aspects depend on contemporaneously related factors. Thus, one needs to consider models that allow for endogenous, dynamic relationships between these two variables. This book is about describing and making informed inferences about such endogenous, dynamic relationships.

This book assumes a certain basic facility with several topics. First, we assume a basic knowledge of matrix algebra and the representation of systems of equations using matrices. Second, we assume a working (and likely good) understanding of general linear regression models. Finally, we assume the reader is familiar with or can consult a basic text on univariate time series analysis such as Harvey (1990), Mills (1991), or Ostrom (1990).

The basic plan of this book is as follows. We first address the philosophical and methodological choices in the specification of multiple time series models. This begins with a treatment of (dynamic) simultaneous equations and univariate time series models. Our goal in this discussion is to illuminate the trade-offs implicit in different specification and identification assumptions. We then discuss how different time series models and simultaneous equation models are related. Finally, we introduce the vector autoregression (VAR) approach to time series as an alternative that can address some of the shortcomings of both simultaneous equation and univariate time series models.

Chapter 2 focuses on the mechanics of simultaneous equation models and VAR models. We address the specification issues and mathematical details of VAR models. This includes a discussion of lag specification, Granger causality (exogeneity) testing, dynamic analysis via impulse responses and decompositions of forecast error variance, and the relationship of VAR to error correction models.

Chapter 3 presents two applications of VAR models. The first is an analysis of the relationship between public mood and aggregate partisan identification in the United States. The second example, which is more extensive and complex, looks at the political and economic determinants of effective corporate tax rates. Here, we look at how multiple time series models are affected by specification decisions about the identification and dynamic properties of the data and theory. We will be making available (via the Internet) the data and code (written in RATS) that we used for these examples so that those trying to learn how to implement these models can see "how" it is done. The data and code are available at the lead author's Web site: http://www.utdallas.edu/~pbrandt.

A concluding appendix offers some guidance on the selection of software for implementing the multiple time series models that we discuss in this book.

—Patrick T. Brandt

MULTIPLE TIME SERIES MODELS

Patrick T. Brandt
University of Texas at Dallas

John T. Williams
University of California, Riverside

1. INTRODUCTION TO MULTIPLE
TIME SERIES MODELS

Many social science data problems are multivariate and dynamic in nature. For example, how is public sentiment about the president's job performance related to the aggregate economic performance of the country? Are arms expenditures by a series of countries related to each other or exogenous? Are the actions directed by country A toward country B related to the actions that country B directs toward country A? How is the percentage of Americans that identify with the major political parties—aggregate partisanship—related to their support for government policies? How are tax rates related to the proportion of political action committees that are organized by business? In each of these examples, it is possible to write down a single equation where one of the variables is the dependent variable and the other is the independent variable. But it is likely that in each of these examples there is simultaneity and that there exists a second equation with the roles of the independent and dependent variables reversed.

For the sample research questions noted above, both the variables are likely to be endogenous. One would expect that the same factors that explain changes in aggregate partisanship are endogenously and dynamically related to public support of government policies. Similarly, changes in tax rates for business are both a cause and a consequence of the lobbying efforts of corporate political action committees. Both these examples (addressed in more detail in Chapter 3) are ones where a researcher might posit two (or more) equations, one for each variable, and allow both the current and the past values of each variable in the model to affect each other.

Most social scientists learn to use regression relatively early in their statistical training. But single-equation regression models ignore the fact that for endogenous, dynamic relationships there is either explicitly or implicitly

1

more than one regression equation. An analyst may choose to continue estimating a single regression and hope that statistical inferences are not too flawed, or he or she might decide to estimate a multiple-equation model using a variety of techniques developed in econometrics (e.g., seemingly unrelated regressions, autoregressive distributed lag [ADL] models, transfer function models). But even this accounts only for specification- and estimation-related issues such as serial correlation and endogeneity. Analysts must also confront the additional complication that the data are measured through time and have time periods as the units of analysis. A researcher looking at these questions could then choose to estimate a single- or multiple-equation model with some time series dynamics. This leads us to consider the relative merits of other avenues of analysis, including vector autoregression, error correction models, and (dynamic) factor analysis. Because time series data are much richer—that is, they contain more information than cross-sectional data—decisions about how to tackle multiple time series problems are crucial.

The need for these dynamic multiple-equation models stems from two very common realities in social science models. First, variables simultaneously influence one another, so both are referred to as endogenous variables. A multiple-equation system usually, but not always, has the same number of endogenous or dependent variables as equations.[1] Although the theoretical interest of an analyst may be on just a single equation, and this equation may be the only one estimated, statistics and econometric theory require that all equations be considered, otherwise inferences can be biased and inefficient. Second, when considering the relationships among multiple dependent variables, the unique or identified relationships for each equation of interest can be made only with reference to the system as a whole. Properly determining these relationships requires that information from all equations be used. Identification requires that there be enough exogenous variables, specified in the correct way, to be able to estimate any or all of the equations in a system. Estimation requires that exogenous variables from the entire system be used to provide the most unbiased and efficient estimates of the relationships among the variables as possible.[2]

In addressing each of these issues, we expect that the dynamic relationships of the variables are of central interest. We would like to know how changes in one of the variables affects the others. It is possible that the relationship among the variables is endogenous in one of two senses. First, changes in one of the variables may have a delayed effect on another (so the effect is through the past values of one variable on the current or contemporaneous values of another). Alternatively, the relationship may be contemporaneous in that changes to the system of equations, known as shocks or innovations, may change both or several variables at the same time. This arises because the shocks to one variable are correlated with the shocks to another variable.

A central concern in translating a theory into an empirically estimable model (i.e., one where we can estimate the parameters and make inferences about them) is that we may not know the *structure* or equation(s) that correctly represents a model. That is, suppose that some (multivariate) probability density $f(y|\beta)$ describes the observed data y in terms of some parameters β. This density would identify a unique structure or set of equations if there is no other set of values β that produce the same density.[3] As social scientists, we do not know for sure if the equations we write down for the specification of a model are the correct ones in many circumstances. A consequence of this is that many disputes over the interpretation of models and their parameters are really based not on the properties of the models per se but rather on the disagreements about the structures or equations used to represent those models.[4] Our goal in this chapter is to illuminate some of the choices faced by social scientists in building theory that accompanies modeling of multiple time series. We will first highlight some of the major choices that analysts must make and then describe the implications of these choices.

Social science theories are built in several stages. First, the researcher must identify the main variables and relationships to be explained with a theory. Here, researchers look at the main theory (or theories) that informs their empirical questions and specifies the relevant variables and relationships to be modeled. Even when there are competing theories present, this stage presents few problems, assuming that relevant time series can be measured for the variables of interest.

Once the main aspects and variables of a theory are determined, the researcher begins the critical phase of selecting the functional form or mathematical structure of the model. It is at this stage that many different models of the same underlying theory or theories begin to emerge. At this stage, we need to make decisions about how theories are translated into equations. This stage will also require that the equations be identified or that there be sufficient restrictions on the equations we specify to ensure that a unique set of parameters can be estimated and interpreted. To do this, we need information about the data, the equations, and the a priori beliefs of the researcher to determine whether or not a model is formally identified.[5]

The third stage of theorizing and model building is fitting the specified model to data and interpreting the results. We note that this step is in one sense noncontroversial because there is wide consensus about how models should be fit and what criteria should be used to evaluate them (unbiasedness, efficiency, minimum mean-squared error, consistency, etc.). A more relevant issue, though, is the determination of the dynamic properties of the specified model from its estimates. Because we are focusing on models of multiple time series, we need to be concerned with methods that can be used to do this.

Finally, once the model has been fit and interpreted, there is often a need to revisit the earlier steps to evaluate the impact of specification decisions and look again at confirmed or disconfirmed aspects of the theory.

The most critical aspects of this model-building process are the specification of the functional form of the system (Stage 2). The remainder of the process, to a large degree, depends on the decisions made in this stage. Models that are specified with missing variables or incorrect dynamics will suffer from the same problems as ordinary least squares (OLS) models— bias and inefficiency. In addition, failing to include a relationship or a factor that is part of the multivariate system can lead to simultaneity biases. Note that these parameter ills then induce problems with interpretation and hypothesis testing.

Standard simultaneous equation models and univariate time series models are commonly proposed to address these types of questions. Although these kinds of models have much to offer, they also have limitations. We next discuss the general decisions that researchers face and highlight some of the trade-offs in the choice of different models. We present the four main approaches that are typically used to model univariate and multiple time series data: autoregressive integrated moving average (ARIMA) models, simultaneous or structural equation (SEQ) systems, error correction models (ECMs), and vector autoregression (VAR). In the remainder of the text, we use simultaneous and structural equation models interchangeably. We discuss how each of the main approaches to modeling dynamic simultaneous relationships forces the researcher to make certain choices about the relationships that may or may not be clearly specified in a theory and the empirical representation and statistical model they specify.

Critical to our presentation is the following: The differences among these methods have less to do with *technique* and more to do with *approach*. All these methods employ some version of linear regression (OLS, generalized least squares, multistage least squares, etc.) or maximum-likelihood methods for estimation. What differ among these methods are the assumptions and building blocks that are the basis for inferences and interpretation.

1.1 Simultaneous Equation Approach

A first approach for building a multiple-equation time series model would be to work in the simultaneous equation (SEQ) paradigm. SEQ models are present in the multiple disciplines of the social sciences. The SEQ paradigm was largely developed by the Cowles Commission in the 1940s and 1950s at Yale. The Commission's early goal was to develop a methodological paradigm for modeling the economy using econometrics. So the researchers there worked to adapt existing econometric methods to the study of large-scale, multiequation models of the economy. In this case, the early Cowles model was an empirical representation of standard Keynesian macroeconomic theory.

Model building with SEQs is based on taking the representation of a single theory or approach and rendering it into a set of equations. Using a single theory to specify the relationships among several variables leads to choices about which variables are exogenous to the system and which are endogenous. The exogenous variables are those that are determined outside the system or are considered fixed (at a point in time or in the past). Those that are determined inside the system and are the dependent variables of the equations are endogenous. The result is a *single* structural system of equations that express the relationships among the variables. The reason there is such a focus on a single theory is that multiple theories may lead to different, typically non-nested specifications of the structural equations (good examples include Zellner, 1971; Zellner & Palm, 2004).

Consider the earlier research question about how aggregate partisanship is related to public support of government policies. A simultaneous equation representation of the relationships among these variables would have two equations, one for each variable. Each endogenous variable would be a function of the other and (possibly) past values of each variable. To estimate such a system, one would need to rewrite the system as a reduced form set of equations where each endogenous variable is a function of predetermined or exogenous values. Unspecified in this modeling approach is how decisions about the number of past values influence the system of equations or how the system of equations will be identified. Typically, (vague) appeals are made to "theory" and hypothesis test results for the inclusion or exclusion of variables.

Several issues arise in constructing SEQ models in this manner. First, alternative theories must be nested within a common structure to be compared. If the models cannot be nested (because of nonlinearity or different specifications), then no single structure can be used to compare different models. Second, the models require that choices be made about the inclusion or exclusion of different variables and lagged values to ensure identification. Two methods are common here: restricting "predetermined" or lagged endogenous variables as exogenous variables and the classification of variables as either endogenous or exogenous. Here, "theory" is used to restrict the parameter space of the model parameters. Often hypothesis tests are used to determine the exclusion of variables, but this then induces pre-test biases in the final models, because the exclusion of variables based on tests typically leads to overconfident results.

As argued by Sims (1980), such exclusion restrictions are often not theoretically justified and are often not well supported by empirical analysis. One consequence of this is that extra lagged values are included or excluded from an SEQ model leading to incorrect dynamic specifications. Even if the models have white noise or serially uncorrelated residuals, the specification search for these models may be incorrect and may imply the wrong dynamic specification, because it has incorrectly restricted the parameter space.

Finally, these models often perform poorly at forecasting and policy analysis. Alternative, simpler models typically will outperform complex, multiple-equation simultaneous equation models.

1.2   ARIMA Approach

Another approach to multiple time series models starts from a time series perspective. One could address multiple time series as a collection of univariate series. In this vein, the researcher would use the standard "Box-Jenkins approach" or ARIMA models for each of the series (Box & Jenkins, 1970). Once the dynamics are known, one could begin to build a model where some of the variables are included as pulse, intervention, or other exogenous effects on the right-hand side of an ARIMA model.

The Box-Jenkins approach is oriented toward forecasting and describing the behavior of a time series (Granger & Newbold, 1986). The general Box-Jenkins approach defines a class of models—in this case ARIMA models—to describe a time series. One then fits a series of ARIMA models to each of the series with the goal being to choose the most parsimonious model with uncorrelated residuals. This approach requires that we designate some of the variables as endogenous and others as exogenous for the fitting of the model. The Box-Jenkins approach is particularly successful at forecasting—in fact, Box-Jenkins-style models will typically outperform SEQ models in forecast performance. The main reason for this is parsimony: The models are built by exploiting the parsimony principle and allowing the data to speak as much as possible.

A Box-Jenkins model for the aggregate partisanship percentage and the percentage of the public supporting the government policies over time would be constructed as follows. Suppose we are most interested in predicting public support. One would first construct a univariate ARIMA model of the public support dynamics. Next, once the model is determined, one would add the aggregate partisanship covariate to see if it improved the fit of the public support model. One would examine various specifications of the partisanship variable (including contemporaneous values and various lags). Hypothesis tests would be used to determine the best specification.

One might reverse the roles of the variables as well and fit a model to the partisanship measure.

There are limitations to this approach when building a multivariate time series model. First, it ignores the fact that some of the variables in the model may help to proxy dynamics in the others. If this is the case, then the suggested procedure may lead to severe overfitting of the data, because the standard Box-Jenkins approach is to filter or explain most of a variable's variance using its own past values. Second, this approach leads us to focus on the dynamics of the variables first rather than on the general relationships in the system. Third, because this is generally done in separate equations—one for each of the variables—we expect that unless the equations are perfectly independent, there will be inefficiency in the estimates. Finally, unless the variables are causally related in a specific way, treating the many variables in separate ARIMA models will lead to inefficient estimates. The reason is that if the residuals for the several variables are contemporaneously correlated (i.e., at the same time), then the estimates will be inefficient. Only when the results of each equation are explicit of the others can we use a sequence of independent equations to model the results.

1.3   Error Correction or LSE Approach

Error correction models are a specialized case of ARIMA regression and simultaneous equation models. They are commonly referred to as the London School of Economics or "LSE" approach because they have been advocated by economists there (Pagan, 1987). The basic building block of an ECM is an autoregressive distributed lag (ADL) specification for two or more variables with provisions for the (possible) long-run relationships among the variables.

The ECM approach differs from ARIMA models in that the long-run relationships—typically, stochastic and deterministic trends—are directly modeled. In ARIMA models, these long-run components, trends or unit roots, are "differenced" from the data to create a stationary data series that can be modeled as an ARIMA process. The ECM approach instead explains the long-run components in two or more data series as a function of each other. The ECM approach uses the long-run component in two or more of the series being modeled to derive a common (stochastic trend) representation that is shared among the series. An ECM uses this common representation to produce a model that has a common long-run component for the variables and a short-run component known as an error correction mechanism that describes how each variable varies or equilibrates around the common long-run component.

The ECMs can be applied to both stationary (mean reverting) and nonstationary (unit root) data. For stationary data, the ECM allows one to estimate a common or equilibrium level for the variables and how each varies around the equilibrium. This model is equivalent to an ADL model, which is an ARIMA model with exogenous variables. For nonstationary or trending data, the ECM modeler starts with a specialized set of data series—two or more series that have unit roots or are integrated to order one.[6] This determination is made using unit root tests (such as the augmented Dickey-Fuller [ADF] test). Once the series are found to be unit roots, a specialized estimation technique is used to estimate both the long- and short-run relationships in the data. For bivariate relationships, a one- or two-step ECM estimation procedure can be used. For multivariate time series (typically with unit roots), the vector ECM (VECM), described in detail in Johansen (1995), is used. The first step in this process is the determination of the common stochastic trend processes in the data. Then, once these long-run trends have been estimated, the short-run dynamics around the long-run trend are estimated using a regression model.

Because the ECM and its multivariate version, the VECM, are based on describing the long- and short-run components of a multivariate time series regression model, researchers can test for a variety of relationships among the common long- and short-run dynamics and how they are related across the various series. For nonstationary data, the representation of the ECM ensures that there is a particular form of "causal" relationship among the series (Engle & Granger, 1987). This causal relationship is known as a "Granger causal" relationship, where the past values of one series must (linearly) predict the current values of the other series. This means that the trend in the two integrated series is "driven" or predicted by the changes in one of the variables. As such, these models are a specialized case of the simultaneous equation models, because they impose and estimate a common trend structure across the series.

Consider again the model of the relationship between aggregate partisanship and the percentage of the public that supports government policies over time. Some argue that these variables are nonstationary or unit roots because they are the sum of a series of accumulated events or shocks that persist (see the discussion and references in Chapter 3 for more details). If this is the case, then an ECM may be appropriate. The ECM can then be used to evaluate how the short- and long-run relationships of these two variables are related and which variable Granger causes the other. The ECM will allow one to apply hypothesis tests to determine the long- and short-run structures of the relationship among the series. The ECM representation of the relationship will provide more information about the dynamics, but we will have to estimate the cointegration relationship and the short-run dynamics.

Inference and model formulation using ECMs, although well developed (e.g., Banerjee, Dolado, Galbraith, & Hendry, 1993), can be quite complicated for nonstationary or unit root data. Many economic variables (e.g., consumption, gross domestic product, government spending) will have unit roots. This is an important reason for considering these models, because they allow one to look at the short- and long-run dynamics of these variables correctly. However, frequentist inference using nonstationary data in these models is complex. This is because of the nonstandard distributions and the complications of computing dynamic analyses when there are unit root variables in the model. This means that hypothesis tests for the presence of error correction relationships and the number of error correction relationships and tests concerning the model parameters often have nonstandard distributions that must be simulated or analyzed using nonstandard test statistic tables (Cromwell et al., 1994; Lutkepohl, 2004). Further, the causal structure of ECMs may be no easier to determine. Sims, Stock, and Watson (1990) note that such inference in models with multiple unit roots is difficult.

1.4   Vector Autoregression Approach

A final approach to modeling multivariate time series is the VAR model. VAR modelers do not assume to know the correct structure of the underlying relationship that generated the multiple time series. Instead, they focus on the underlying correlation and dynamic structure of the time series.

The VAR approach starts by focusing on the interrelated dynamics of the series. It asks the following questions (in contrast to the SEQ approach):

1. What is to say that some lagged variables would not be in each equation? Does restricting the dynamics for identification make sense?
2. What impacts do each of the variables have on each other over time?
3. If a variable affects one equation in the system of equations, what is to say that it does (or does not) affect another?
4. Are rational expectations—the idea that a variable is best predicted by its immediate past value plus a random component—present? In this case, the past is of little predictive value, and policy makers and analysts are interested in how the random components—innovations or policy shocks—are translated into outcomes. In this framework, the shocks are themselves exogenous variables.

These are all critiques of the standard (i.e., Cowles Commission) approach to simultaneous equation models. The main difference in the VAR approach is that it is built on creating a complete dynamic specification of the series in a system of equations. The basic idea of VAR modeling is built on the insights of the Wold decomposition theorem (Hamilton, 1994, pp. 108–109; Wold, 1954). Wold showed that every dynamic time series could be partitioned into a set of deterministic and stochastic components.

All these critiques point toward understanding dynamics. In response to these critiques, Sims (1972, 1980) pioneered the VAR methodology, building on the idea of dynamic decomposition of the variables in the system. Sims rejected the use of standard simultaneous equation models for three reasons:

1. Identification restrictions on parameters used in SEQ models are typically not based on theory and thus may lead to incorrect conclusions about the structure of the models and the estimates.
2. SEQ models are often based on tenuous assumptions about the exogeneity and endogeneity of the variables. Because the true lag lengths of the variables are not known a priori, identification is then based on possibly specious assumptions about exogeneity. The formal identification of a dynamic simultaneous equation model requires that the exact true lag length be known for each variable; otherwise, identification assumptions may not hold (Hatanaka, 1975).
3. If the variables in the model are themselves policy projections, additional identification problems will be present because of temporal restrictions. This is the rational expectations critique: Models are typically treated as though ceteris paribus claims will be true. In fact, if they are not, then we need to be able to assess the probabilistic implications of different paths for the variables.

The Sims-proposed method for addressing the tenuous identification problems of the SEQ approach is to focus on the dynamic specification of the reduced form model. This is in contrast to the SEQ approach, which focuses on the identification choices in the model specification. Sims's approach is to ensure that the modeling approach to multiple time series provides a complete characterization of the dynamics of several series. This is done using a multivariate autoregressive model to account for the dynamics of all the variables.[7]

The VAR model proposed by Sims is a multivariate autoregressive model where each variable is regressed on its past values and the past values of the other variables in the system. Model building in VAR models then depends on the selection of the appropriate variables (based on theory). The specification of the dynamic structure proceeds based on testing for the appropriate lag length using the sample data. Sims (1980) argues that one of the critical contributions of the VAR approach is that it can serve to define the "battleground" of empirical debates about multiple time series data. It does this by providing a model of the dynamic and empirical regularities of a set of related time series. From this point, one is able to refine and develop the empirical model to evaluate theoretical debates.

The logic of the VAR approach can be applied to the aggregate partisanship and public support of government policies example. Instead of starting with a set of structural (e.g., SEQ), dynamic (e.g., ARIMA or ECM), or a priori causal relationships (e.g., ECM) among the variables, a VAR model begins by assuming that the reduced form dynamics are of central interest. Thus, rather than impose possibly structural or dynamic restrictions on the relationships among the two series, a VAR would have two equations, one for each variable. Each variable would be regressed on its past values and the past values of the other variable. The resulting residuals (after checking for serial correlation) would be exogenous shocks or innovations. One could look at the responses of each equation to see how these "surprises" in each variable affect the observed system. After accounting for these (historical) dynamics, one could engage in inferences about the Granger causal relationships between the two variables and try to determine the endogenous structure and dynamics of the two series.

Building multivariate time series models according to the VAR methodology does not then depend on a single theory. Instead, multiple theories can be compared explicitly and evaluated (using hypothesis testing) without the identification assumptions that would be made in the specification of alternative simultaneous equation models. Because the variables in the VAR model are not segmented a priori into endogenous and exogenous variables, we are less likely to violate the model specification and incorrectly induce simultaneity biases by incorrectly specifying a variable as exogenous when it is really endogenous.

The key distinction between VAR and SEQ models is the treatment of identification assumptions. In the SEQ model, these are taken as fixed, invariant, and specified by theory. In the VAR approach, such zero-order restrictions (e.g., excluding variables from some equations or omitting some lagged values of some variables from some equations) are seen as unlikely to be true. Thus, in an effort to eliminate biases from these incorrect restrictions, VAR models are able to consider trading off these biases for some inefficiency. The biases in the SEQ model estimates are the result of omitting lagged values that should be included in the models. Under Sims's logic, often some lags of some variables are (incorrectly) excluded to identify the SEQ model. These incorrect restrictions, which lead to the omission of relevant lagged variables, produce omitted variable bias. The solution to mitigating this bias is to include all possible lags (which may be more than necessary). In this case, the goal is to reduce bias at the cost of efficiency.

The key identification assumption in a VAR model is how the contemporaneous effects of each variable are related to each other. Because the VAR model is specified in terms of the lagged values of the variables in the system on each other, identification concerns the specification of the residual or contemporaneous covariance matrix of the residuals alone. The benefit of this is that it allows one to separate the interpretation of the model dynamics from the identification. This allows researchers to explicitly look at how identification decisions are related to the path of the variables' dynamics.

VAR modelers also have a different conception of the interplay of data and models. The goal of a VAR model is to provide a probability model of the dynamics and correlations among the data (Sims, 1980). Thus, VAR models are considered best when based on a simple, unbiased specification that accounts for the uncertainty about the dynamics and the model. To do this, pretest biases must be avoided (Pagan, 1987). Thus, unlike the "specify-estimate-test-respecify" logic of classical approaches, SEQ models, and ARIMA models, VAR models employ few hypothesis tests to justify their specification. This leads to a less biased representation of the model and its dynamics rather than the false sense of precision that can accompany other modeling strategies. That is, once we have entered this cycle of specification testing, the resulting inferences are a function of the test procedure and are less certain than the reported test statistics and associated levels of significance or reported P-values would lead us to believe.

1.5   Comparison and Summary

This brief review of possible approaches to multiple time series models has been intended to connect other approaches to the VAR methodology (see, e.g., Pagan, 1987; Sims, 1996). The ARIMA, ECM, and SEQ methods are special cases of the more general VAR methodology. Freeman, Williams, and Lin (1989) presented a basic comparison of the VAR and SEQ approaches. Here, we have discussed these and other models with the goal of comparing how dynamics are modeled and how inferences are made.

Table 1.1 presents a summary of each of these methods, which extends the summary initially presented in Freeman, Williams, and Lin (1989). The table shows the main methodological differences in the specification of time series models.

The critical point is that VAR models are a generalization of the other approaches. Each of the other modeling approaches focuses on some feature of time series data that *may be* true in practice. However, from the standpoint of model formation and theory testing, the more general approach is a VAR model.

TABLE 1.1
Comparison of Time Series Modeling Approaches

	ARIMA	Error Correction	SEQ	VAR
Model Building				
Specification	Single theory focused on univariate series	Long- and short-run trends and dynamics based on results of tests for cointegration and unit roots	Single theory with assumptions about endogeneity and exogeneity	Recognition of multiple theories by including variables as endogenous
Estimation	Maximum likelihood; OLS	Johansen procedure; one- or two-step procedure	Higher-order OLS and maximum-likelihood methods; corrections for heteroscedasticity and serial correlation; tests for overidentification and orthogonality	OLS and tests for lag length
Methodological Conventions				
Hypothesis testing	Analysis of individual coefficients	Tests of cointegrating relationships; short-term dynamics	Analysis of individual coefficients; goodness of fit	Analysis of significance of blocks of coefficients; tests for exogeneity
Dynamic analysis	Dynamic multipliers; intervention analysis	Analysis of cointegrating vector; impulse responses	Simulation; deduction of model dynamics	Forecasts, model projections, decomposition of forecast error variation, impulse responses

Why, then, do we advocate the VAR approach? First, we are not trying to rule out structural models. In fact, such structural models can and will do a good job in some cases when the restrictions in those models are true. They will provide better inferences, summaries of dynamics, representations of relationships, and other measures. Second, using a step-by-step approach to structural model building *that is explicit in its treatment of assumptions about dynamics* will be to produce a better model based on the parsimony principle.

When are these structural equation models a poor choice, and when should they be replaced by VAR models? There are three situations, all of which are when these models fail. First, unless we know or test the precise structure of the relationships among the variables in our models, SEQ models will be misspecified. Second, for policy or counterfactual analysis, unless the models are correct, we may make incorrect inferences. Finally, if one of our goals is to characterize uncertainty and dynamics, then VAR models will typically be superior because they are less likely to be overly precise via ad hoc pretesting of the models under consideration.

Finally, there are times when one would prefer to use an ECM. These are when one wants to isolate the long- and short-run behaviors of several series simultaneously, or when trending or unit root variables are present in a multiple time series model. In these cases, one is actually estimating a VAR model with a set of restrictions or assumptions (which can be tested) about how the long-run behavior of the multiple time series model evolves. Our contention in this book is that to understand and apply the ECM and VECM models, one should have a solid basis in the more general, unrestricted VAR approach. We return to the relationship between VAR and ECM models in the next chapter.

The next chapter outlines the mathematical details of a VAR model. We then discuss how this model is used for inference about the relationships in multivariate time series data.

2. BASIC VECTOR AUTOREGRESSION MODELS

Vector autoregression (VAR) models are not a statistical technique or methodology. Rather, VAR models are an approach to modeling dynamics among a set of (endogenous) variables. This approach focuses on the dynamics of multiple time series and typically employs multiple regression and multivariate seemingly unrelated regression models. The central focus is on the data and their dynamics. A central tenet of VAR models is the idea that restrictions on the data and parameters in the model should be viewed skeptically.

How skeptically? Consider some hypothetical time series data that have a rich dynamic and correlated structure. Consider a "view" of these data: With perfect knowledge, you can see these data and know their rich dynamics. Now consider closing one eye. With one eye closed, you see only part of the dynamics of the data (the left or the right depending on which one you closed) and may lack the full perception of the depth of these data. VAR modeling is an effort to force you to keep your eyes open rather than incorrectly closing your eyes or occluding your vision with incorrect assumptions.

What, then, is a VAR model? Simply, it is an interdependent reduced form dynamic model. For each endogenous variable in the system of equations, one constructs an equation that makes these variables functions of their own past values and the past values of all other endogenous variables. Typically, this will be done with the same number of lags or past values of each variable in each equation. Additional exogenous or control variables can easily be included as additional regressors.

In this chapter, we present the mathematical details needed to specify and interpret a basic VAR model. We assume that the reader has no more knowledge than a basic understanding of linear regression models and matrix algebra. From this base, we develop the specific aspects of a VAR model.

This chapter has two main purposes. The first is to show how the VAR model is related to other well-known models such as structural equation models. Second, we present the basic terminology and techniques used to specify, estimate, and interpret VAR models. Chapter 3 then presents two examples based on our discussion.

This chapter is organized as follows: First, we present a generic dynamic simultaneous equation model and a VAR model and show how the VAR approach can help us "do no harm" in understanding the data dynamics. Next, we discuss the main specification and inferential decisions that are part of a VAR model. We then discuss in detail the selection of lag lengths, estimation, and inference in the VAR model. We pay particular attention to the presentation of dynamic response analysis via impulse responses and decompositions of forecast error variance methods. Finally, we discuss several other topics in the specification of VAR models that are often overlooked, but are considered standard practice.

2.1   Dynamic Structural Equation Models

Chapter 1 outlined the basic contributions and developments of dynamic structural equation models in econometrics and the social sciences. Here, we are looking at the structure of these models. To fix ideas, consider a basic dynamic structural equation model for a set of two endogenous variables Y_t and Z_t. Each of these variables is observed at times $1, \ldots, T$. The lagged or past values of Y_t and Z_t are denoted by the lagged subscripts $Y_{t-\ell}$ and $Z_{t-\ell}$, respectively, where $\ell = 1, 2, \ldots$ denotes the lag length for the observation ℓ periods before t.

A dynamic simultaneous equations system for these variables is

$$Y_t = \alpha Z_t + \gamma_{11} Y_{t-1} + \gamma_{12} Z_{t-1} + u_{1t}, \tag{2.1}$$

$$Z_t = \theta Y_t + \gamma_{21} Y_{t-1} + \gamma_{22} Z_{t-1} + u_{2t}, \tag{2.2}$$

where

$$u_{it} \sim N\left(0, \begin{pmatrix} \sigma_{11} & \sigma_{12} \\ \sigma_{12} & \sigma_{22} \end{pmatrix}\right).$$

This is a simultaneous system because all the relationships (equations) in the model are necessary to determine the value of at least one of the endogenous variables in the model. In other words, we see the contemporaneous values of each variable in each equation—Z_t on the right-hand side of Equation 2.1, and Y_t on the right-hand side of Equation 2.2. The simultaneity in the model results from the fact that each variable depends on the contemporaneous value of the other variables in the model. The dynamics arise from the lagged values.

We refer to the variables Y_t and Z_t as the endogenous variables. Those that have already been determined in the model at time t (i.e., Y_{t-1} and Z_{t-1}) are referred to as *predetermined* and can be either exogenous variables or lagged endogenous variables whose values are already known. Note that this model contains both a summary of the variables related to each of the endogenous variables and the temporal relationships among the variables.

The model in Equations 2.1 and 2.2 is referred to as the *structural equation form* of the model. Note that to be able to estimate this system of equations, we must substitute one of the equations into the other, because at least one of the equations is necessary to determine the other (by definition). Also, we cannot separate the two equations and estimate each by ordinary least squares (OLS): Ignoring one of the equations means that one is ignoring the simultaneity among the variables. This will lead to simultaneity bias in the regression estimates. In addition, the system for the two variables is dynamic, because the values of each of the variables at time t depend on the values of the variables at time $t-1$. This is an autoregressive process of order one or "AR(1) process" for each of the variables.

To produce a set of equations that can be estimated consistently, we derive the *reduced form*. This is done by substituting Equation 2.2 for Z_t into Equation 2.1 for Y_t and solving for Y_t. This yields the following equation for Y_t:

$$Y_t = \alpha[\theta Y_t + \gamma_{21} Y_{t-1} + \gamma_{22} Z_{t-1} + u_{2t}]$$
$$+ \gamma_{11} Y_{t-1} + \gamma_{12} Z_{t-1} + u_{1t},$$
$$= \alpha\theta Y_t + \alpha\gamma_{21} Y_{t-1} + \alpha\gamma_{22} Z_{t-1} + \alpha u_{2t} + \gamma_{11} Y_{t-1}$$
$$+ \gamma_{12} Z_{t-1} + u_{1t},$$
$$Y_t(1-\alpha\theta) = (\alpha\gamma_{21} + \gamma_{11}) Y_{t-1} + (\alpha\gamma_{22} + \gamma_{12}) Z_{t-1} + \alpha u_{2t} + u_{1t},$$
$$Y_t = \frac{(\alpha\gamma_{21} + \gamma_{11})}{(1-\alpha\theta)} Y_{t-1} + \frac{(\alpha\gamma_{22} + \gamma_{12})}{(1-\alpha\theta)} Z_{t-1} + \frac{\alpha u_{2t} + u_{1t}}{(1-\alpha\theta)}.$$

The reduced form equation for Z_t is found analogously:

$$Z_t = \theta[\alpha Z_t + \gamma_{11}Y_{t-1} + \gamma_{12}Z_{t-1} + u_{1t}]$$
$$+ \gamma_{21}Y_{t-1} + \gamma_{22}Z_{t-1} + u_{2t},$$
$$= \theta\alpha Z_t + \theta\gamma_{11}Y_{t-1} + \theta\gamma_{12}Z_{t-1} + \theta u_{1t} + \gamma_{21}Y_{t-1}$$
$$+ \gamma_{22}Z_{t-1} + u_{2t},$$
$$Z_t(1 - \theta\alpha) = (\theta\gamma_{11} + \gamma_{21})Y_{t-1} + (\theta\gamma_{12} + \gamma_{22})Z_{t-1} + \theta u_{1t} + u_{2t},$$
$$Z_t = \frac{(\theta\gamma_{11} + \gamma_{21})}{(1 - \theta\alpha)}Y_{t-1} + \frac{(\theta\gamma_{12} + \gamma_{22})}{(1 - \theta\alpha)}Z_{t-1} + \frac{\theta u_{1t} + u_{2t}}{(1 - \theta\alpha)}.$$

This solved system is known as the *reduced form* of the model. These equations show explicitly how the endogenous variables are related to the predetermined variables. The reduced form can be written more compactly as

$$Y_t = \Pi_{11}Y_{t-1} + \Pi_{12}Z_{t-1} + \epsilon_{1t}, \tag{2.3}$$

$$Z_t = \Pi_{21}Y_{t-1} + \Pi_{22}Z_{t-1} + \epsilon_{2t}, \tag{2.4}$$

where Π_{ij} are the coefficients of the earlier reduced form equations in equation i for variable j. This reduced form system for Y_t and Z_t can be estimated consistently by OLS methods. Yet the estimates produced by the OLS estimation of the reduced form, Π_{ij}, are not the parameters of interest for the theoretically specified structural model in Equations 2.1 and 2.2. We are really interested in the structural model coefficients, which must then be derived from the reduced form. Although we have six structural model parameters of interest, the reduced form model generates only four parameters. However, the recovery of the six-parameter model from the four-parameter model is difficult. It presents the researcher with a choice, because in this case, we must make an assumption to identify the model. Which assumption we make affects our inferences about the parameters in the structural model and our description of the dynamics.

To see why this is the case, suppose one wants to recover a consistent estimate of α in Equation 2.1 from the reduced form parameters. A consistent estimate of α requires that there is no feedback from Equation 2.2 for Z_t into Equation 2.1 that would correlate with their parameters. Mathematically, this condition is $\theta = 0$ so that $E[Z_t u_{1t}]$ would then equal zero. In addition, to know if an OLS estimate of the parameter α is consistent, we also need to know if the estimate of $\Pi_{21} = 0$, so that there is no feedback in the system of equations that would invalidate the OLS estimates of α via the past values of this variable. Knowing the reduced form parameter Π_{21}, however, does not tell us whether the requirement for no feedback, $\theta = 0$, is true. This is the condition we need for the α estimate to be consistent.

Why is this the case? Consider three scenarios or identification assumptions that could be made to determine the relationship between θ and Π_{21}:

1. $\gamma_{21} \neq 0$, but $\theta = 0$,
2. $\gamma_{21} = -\theta\gamma_{11}$, so that $\Pi_{21} = 0$, but $\theta \neq 0$, or
3. $\gamma_{21} = 0$ and $\gamma_{11} = 0$, but $\theta \neq 0$.

In the first case, we assume that only the past values of Y_t can be used to predict Z_t. In this case, there would be no feedback, but the value of Π_{21} would be γ_{21}. In the second case, we would be assuming that the coefficients for the past and the present values of Y_t canceled each other out proportionately and were not useful for predicting Y_t or Z_t. But in this case, $\Pi_{21} = 0$ and θ would not be zero—so the estimates of α would be inconsistent. Finally, in the last case, we see that θ can be nonzero and still generate a reduced form coefficient that is zero.

What this tells us is that the decisions about the relationships of the reduced form parameters to the structural parameters are critical for the interpretation and statistical validity of the model estimates. These decisions about how to recover the structural parameters from the reduced form also place restrictions on the dynamics of the simultaneous equation system. In the first case, we are assuming (possibly incorrectly) that only the past values of Y_t matter for predicting Z_t. In the second case, the dynamic restriction is that it is the contemporaneous correlation of the two variables that describes their dynamics because the impact of the past values of Y_t cancels out across the two equations. In the last case, the past values of Y_t have no predictive value in either equation—all of the explanatory power of Y_t for Z_t is in terms of the contemporaneous values alone. All of these possible identification restrictions place some limitation on the possible feedback or dynamics in the system of equations.

These are very different dynamic models of the relationship among Y_t and Z_t. And the choice is driven not by any theory or set of theories, but rather by the need to achieve identification for consistent estimation. But it is not necessary to do this. Consider the alternative interpretations discussed in Chapter 1. We can analyze just the reduced form—without the possibly incorrect restrictions necessary to identify the structural model. In this case, we focus on the dynamic relationships of the variables and allow for a myriad of possible contemporaneous relationships.

2.2   Reduced Form Vector Autoregressions

As an alternative to the possibly incorrect identifying assumptions of the simultaneous equation system in Section 2.1, Sims (1972, 1980) proposed working directly with the reduced form model. Because in time series analysis we are concerned with the *dynamics* of the system of equations, this is not problematic. In fact, using the reduced form we will still be able to investigate various assumptions about the nature of the identification assumptions *after assessing the dynamics of the system of variables we are interested in modeling.*

In the VAR model for this dynamic system, we write the system of endogenous variables as a function of their predetermined variable values and the predetermined values of the other variables in the system. The VAR is, then, a system of unrestricted reduced form equations. To write equations for a system of m endogenous variables, let y_{it} be the ith variable where $i = 1, 2, \ldots, m$ and t is the time index. A scalar version of the reduced form VAR would include the following equations:

$$y_{1t} = \beta_{10} + \beta_{11} y_{1, t-1} + \cdots + \beta_{1, p} y_{1, t-p} + \cdots$$
$$+ \beta_{1, m} y_{m, t-1} + \cdots + \beta_{1, mp} y_{m, t-p} + e_{1t},$$
$$y_{2t} = \beta_{20} + \beta_{21} y_{1, t-1} + \cdots + \beta_{2, p} y_{1, t-p} + \cdots$$
$$+ \beta_{2, m} y_{m, t-1} + \cdots + \beta_{2, mp} y_{m, t-p} + e_{2t}, \qquad (2.5)$$
$$\vdots$$
$$y_{mt} = \beta_{m0} + \beta_{m1} y_{1, t-1} + \cdots + \beta_{m, p} y_{1, t-p} + \cdots$$
$$+ \beta_{m, m} y_{m, t-1} + \cdots + \beta_{m, mp} y_{m, t-p} + e_{mt}.$$

For each of the m endogenous variables, we regress that variable on its p lagged values and the p lagged values of each of the other endogenous variables. Thus, each equation contains $mp + 1$ regression coefficients for a total of $m(mp + 1) = m^2(p + m)$ coefficients for the entire system. The model is completed by making an assumption about the distribution of the residuals. In this case, we assume that the residuals are jointly distributed normally, or $e_t \sim N(0, \Sigma \otimes I)$, where $e_t = (e_{1t}, \ldots, e_{mt})$, Σ is the $m \times m$ matrix of the covariance of the residuals, I is a $T \times T$ identity matrix, and $\otimes$ is the Kronecker product operator for two matrices.[1] This model is a multivariate regression model where all the endogenous variables have been put on the left-hand side of the equations and all the lagged or predetermined variables are included on the right-hand sides of the regression equation.

An often seen alternative representation for the reduced form VAR in Equation 2.5 is as a system of equations represented using matrices. Here, the m endogenous variables at time t are in a $(1 \times m)$ vector $y_t = (y_{1t}, y_{2t}, \ldots, y_{mt})$ and the β coefficients are collected into a set of matrices B_ℓ:

$$y_t = c + \sum_{\ell=1}^{p} y_{t-\ell} B_\ell + e_t, \qquad (2.6)$$

where c is the vector of intercepts (β_{i0}), $y_{t-\ell}$ is the $1 \times m$ vector of the ℓth lagged variables, $m \times m$ matrices B_ℓ are the coefficients for the ℓth lag (β_{i1} to $\beta_{m,mp}$ collected by lag), and e_t is a $1 \times m$ vector of the residuals.

A final representation of Equation 2.5 can be used to show that this is a special case of a multivariate regression model. The model can be rewritten in a matrix notation by stacking all the observations from $1, \ldots, T$,

$$Y = XB + e, \qquad (2.7)$$

with

$$Y = \begin{pmatrix} y_{11} & \cdots & y_{m1} \\ \vdots & \ddots & \vdots \\ y_{1T} & \cdots & y_{mT} \end{pmatrix}, \quad X = \begin{pmatrix} y_{1,t-1} & \cdots & y_{mp,t-p} \\ \vdots & \ddots & \vdots \\ y_{1,T-1} & \cdots & y_{mp,T-p} \end{pmatrix},$$

where Y is a $T \times m$ matrix of the endogenous variables for times $t = 1, \ldots, T$ and X is a $T \times (mp + 1)$ matrix of the lagged endogenous variables. The matrix B is an $(mp + 1) \times m$ stacked matrix of the components of the autoregressive matrices, B_ℓ, for $\ell = 1, \ldots, p$.

Note that these reduced form VAR models have the same form as the reduced form of the SEQ model derived in Equation 2.3. In Equations 2.3, 2.5, and 2.6, the values of the endogenous variables at time t are the dependent variables and the right-hand side variables of the equations are the lags. Here, there are no arbitrary restrictions on the number of lags included because the same number of lagged values are included for each variable in each equation. In the next section, we show more formally how VARs are a reduced form representation of a general dynamic simultaneous equation model. We then discuss how VAR models can be used to understand dynamics, how they are estimated, and how inferences can be made using them. In subsequent sections, the VAR models are mainly presented using the notation in Equations 2.6 and 2.7.

2.3   Relationship of a Dynamic Structural Equation Model to a Vector Autoregression Model

In this section, we discuss how VAR models are reduced form representations of general dynamic simultaneous equation models. To do this, we first define a more general version of the dynamic simultaneous equation model that was discussed in Section 2.1. Suppose we have the following dynamic system of equations in vector form:

$$y_t A_0 = d + y_{t-1} A_1 + y_{t-2} A_2 + \cdots + y_{t-p} A_p + u_t, \qquad (2.8)$$

where the row vectors for the parameters and data are defined (similar to Equation 2.6) as

$$y_t = (y_{1t}, y_{2t}, \ldots, y_{mt}),$$
$$u_t = (u_{1t}, u_{2t}, \ldots, u_{mt}),$$
$$d = (d_1, \ldots, d_m), \quad \text{and} \quad u_t \sim N(0, I).$$

Here, the residuals are assumed to be mean zero with an $m \times m$ identity covariance matrix. The A_i matrices are $m \times m$ matrices that define the impacts of the lagged values of the endogenous variables in the system. The A_0 matrix defines the contemporaneous relationships among the endogenous variables. To ensure that the system is identified, we require that A_0 is full rank and invertible; that is, A_0^{-1} exists.

If we postmultiply both sides of Equation 2.8 by A_0^{-1}, then the result is

$$y_t = c + y_{t-1} B_1 + y_{t-2} B_2 + \cdots + y_{t-p} B_p + e_t, \qquad (2.9)$$

where

$$c = d A_0^{-1}, \quad B_i = A_i A_0^{-1}, \quad i = 1, 2, \ldots, p, \quad e_t = u_t A_0^{-1}.$$

Equation 2.9 is a reduced form representation of Equation 2.8. This reduced form representation also has the same form as the general VAR model in Equation 2.6—the contemporaneous values of the endogenous variables regressed on their own past values and the past values of the other variables in the system.

The *contemporaneous relationships* in the reduced form VAR model are then included in the parameterization of the residual covariance. The orthogonalization or identification of the contemporaneous relationships is accomplished through the parameterization of the elements in the matrix A_0. To see how the contemporaneous relationships of the structural model become part of the reduced form residuals, we start with the reduced form residuals e_t and compute their covariance matrix of the reduced form residuals, Σ, as a function of the contemporaneous structure A_0 as follows:

$$\Sigma = V[e_t] = E[e_t' e_t]$$
$$= E[A_0^{-1'} u_t' u_t A_0^{-1}]$$
$$= A_0^{-1'} E[u_t' u_t] A_0^{-1}$$

$$= A_0^{-1\prime} I A_0^{-1}$$
$$= A_0^{-1\prime} A_0^{-1}.$$

This is an important conclusion about VAR models: Restrictions on the contemporaneous relationships—what would be the A_0 matrix in simultaneous equation models—are reflected in the residual covariance relationships in VAR models.

Why adopt this approach where the contemporaneous relationships—which are the source of the need for exclusion restrictions for identification in SEQ models—are included in the variance of the residuals? As noted in Chapter 1 and in Section 2.2, this is the *unrestricted reduced form*—the reduced form containing all the variables of interest without imposing possibly incorrect identification assumptions, but allowing for the most unrestricted representation of the dynamics. We can then use the estimated version of this model to evaluate both structural and dynamic specification decisions without a priori biasing our inferences about the parameters of interest.

2.4 Working With This Model

How, then, do we work with this model? That is, because it is not tied to a specific structural model, what purposes does it serve? There are three main things that analysts typically investigate with VAR models: the causal effects of the endogenous variables on each other, the dynamic impact of changes in one variable on the others in the model, and the amount of the variance in each variable that can be attributed to changes in each variable itself and the other variables in the system of equations.

The first of these, assessing causal relationships, is done using the idea of Granger causality. To determine Granger causality, a hypothesis test is used to determine whether one variable is statistically useful for predicting another. If it is, then we can start constructing causal orderings based on the data relationships that can be used to achieve more specific identification assumptions about a system of equations. The value of these tests lies in the realization that many substantive social science hypotheses can be cast as asking about the exogeneity of one variable with respect to another. Granger causality provides a way to relate this exogeneity of a variable X to a variable Y to the predictive value of including variable X in equations for variable Y.

The second, assessing the dynamic impacts, is known in single-equation time series analysis as measuring the impact multipliers—the short- and long-run impacts of changes in the right-hand-side variables. The multiple-equation analog to this is known as impulse response function (IRF) or moving average response (MAR) analysis. These are multivariate dynamic multipliers that can be used to show whether there are dynamic, causal relationships among the variables. Impulse responses are found by inverting the system of VAR equations to find the moving average representation of the model (subject to identification assumptions). The reason for switching to the moving average representation is that we can then look at the impact of (exogenous) shocks to the equations in the system and trace out the response of the system to this shock. One can use this moving average representation of the model to analyze the implications of different identification assumptions about the contemporaneous correlations of the endogenous variables.

The final way of assessing a VAR model looks at determining how much of the variation of each of the variables is due to the dynamic changes in the other variables. This is known as innovation accounting or decomposition of the forecast error variance. With this method, we ask how much of the forecast error in the variables is due to variable i and how much can be attributed to variable j. This allows us to see how innovations or changes in one variable lead to change in another. The more of a variable's forecast errors that are attributed to another variable, the more important that second variable is for predicting or explaining the first variable. Here, we ask how much of the forecast dynamics of variables evolves from the contemporaneous relationships among the variables and how much emerges from the dynamics of the system of equations.

The goal of all these possible interpretations of VAR analysis is to understand the dynamic impacts of changes in variables based on the reduced form specification. Instead of focusing on the specific parameters in a structural model (which may or may not be correctly identified), we focus on tracing out the dynamic properties of the models. All the methods discussed above analyze how a set of related time series variables are related to each other over time.

Before turning to these methods of interpreting VAR models, we address details of VAR model specification and estimation. We then return to the topics of Granger causality, impulse response analysis, and innovation accounting.

2.5   Specification and Analysis of VAR Models

The initial specification of a VAR system of equations requires the researcher to confront a slightly different set of issues from those in standard autoregressive integrated moving average (ARIMA) or Box-Jenkins univariate time series analysis. First, the multivariate nature of the data complicates and basically negates the value of using diagnostic tools such as autocorrelation functions for specification searching. Second, the estimation of the reduced form VAR model requires specification of the lag length p. Third, the interpretation of the dynamics by looking at shocks to the equations requires that we identify the contemporaneous relationships among the variables in the model. Fourth, the specification of the system determines the method of estimation for the VAR model. The subsequent sections of this chapter address these issues.

2.5.1 Estimation of VAR

For the unrestricted VAR model, the maximum-likelihood estimator for Equation 2.7 is

$$B = (X'X)^{-1}X'Y. \tag{2.10}$$

Note that B is an $(mp+1) \times m$ matrix of regression coefficients, where the jth column contains the regression coefficients for the jth variable. Because the error covariance of the residuals is assumed to be block diagonal (see above), this is a seemingly unrelated regression model. As a result, we do not need to estimate all m equations at once: We can instead use equation-by-equation OLS to obtain consistent estimates.[2] In these cases, where there are no restrictions placed on the VAR coefficients, estimation is done by a series of m OLS regressions, one for each dependent variable in the VAR.

The covariance matrix of the residuals (discussed in Section 2.3) can be estimated from the sample residuals

$$\hat{\Sigma} = \frac{1}{T} \sum_{t=1}^{T} \hat{e}_t' \hat{e}_t, \tag{2.11}$$

where $\hat{e}_t$ is the $1 \times m$ matrix of residuals from the multivariate regression in Equation 2.6. This is the sample error covariance matrix over all the observations.

2.5.2 Lag Length Specification

The standard results about the properties of the VAR coefficients and VAR estimation (e.g., Hamilton, 1994, chap. 11) depend on the lag length of the VAR. In the earlier discussion, we have assumed this is known. However, we typically must evaluate a series of models with different lag lengths to select the appropriate lag length p.

There are two main approaches used for selecting or testing for lag length in VAR models. The first consists of rules of thumb based on the periodicity of the data and past experience (particularly with cyclical economic data). The second is based on formal hypothesis tests.

2.5.2.1 Rules of Thumb for VAR Lag Length Specification

As a rule of thumb, VAR models typically include enough lags to capture the full cycle of the data. So for monthly data, this means that there is a minimum of 12 lags. More typically, we expect that there is some seasonality that carried over from year to year and across the months, so often lag lengths of 13–15 months are used. For quarterly data, one often uses 6 lags. This captures the cyclical components in the year and any residual seasonal components in most cases. To ensure robustness, one should probably evaluate models with up to 8 or 10 lags using the tests discussed below. For both the monthly and the quarterly data, these lag lengths will capture most major forms of seasonality. This is important because for even deseasonalized data, there are likely to be residual seasonal patterns in the data that need to be modeled.[3]

A final rule of thumb for selecting the lag length is that it should not lead to using more than one quarter of the degrees of freedom available for any equation. Thus, if the time series of interest have 120 time periods, the number of lags should be no more than $mp + 1 < T$, where m is the number of endogenous variables, p is the lag length, and T is the total number of observations. So, if there are three variables in the system, one would want to look for lag lengths that satisfy $3p + 1 < 120$ or $p < \frac{119}{3}$ or about 40 lags maximum. The reason for this limitation is twofold. First, using too many degrees of freedom in the VAR will lead to relatively inefficient estimates. Second, using too many lags may make it impossible to estimate the coefficients because the OLS estimates cannot be computed.[4]

Although these rules are general, they often provide only a starting point for determining the appropriate lag length. For data that are not quarterly or monthly, they provide no guidance. In addition, we can use the hypothesis tests described below to more formally evaluate the different specifications for lag lengths of p and $p - 1$.

2.5.2.2 Tests for VAR Lag Length Specification

Two classical test statistics can be used to evaluate the lag length of a VAR model. The first is based on a likelihood ratio test. A likelihood ratio test compares the maximum value of the likelihood achieved for a model with p lags to a model with $p - 1$ lags. The likelihood function for the VAR model can be written as

$$L(\hat{\Sigma}, \boldsymbol{B}, p) = -\frac{Tm}{2}\log(2\pi) + \frac{T}{2}\log|\hat{\Sigma}^{-1}| - \frac{Tm}{2}, \qquad (2.12)$$

where $\hat{\Sigma}^{-1}$ is the matrix inverse of the estimated error covariance matrix (see Equation 2.11) and $\log|\hat{\Sigma}^{-1}|$ is the logarithm of the determinant of

$\hat{\Sigma}^{-1}$, the inverse of the error covariance matrix. Note that the dimensions of B in the likelihood function depend on the lag length p. Thus, we can be explicit and write $L(\hat{\Sigma}, B, p)$ to define the likelihood function.

Formally, we can construct a likelihood ratio test or χ^2 test for models of different lag lengths by comparing the likelihoods for two models with different lag lengths. It is important to note that this χ^2 test is useful for determining whether or not the VAR model accounts for all of the dynamics. The test, however, is only accurate asymptotically (as $T \to \infty$). The null and alternative hypotheses for these model are

H_0: **Null model** The VAR model has $p = p_0$ lags.
H_A: **Alternative model** The VAR model has $p = p_1$ lags where $p_1 > p_0$.

The likelihood ratio test comparing a model with $p_1 > p_0$ lags can be written as

$$(T - 1 - mp_1)\left(\log|\widehat{\Sigma_0}| - \log|\widehat{\Sigma_1}|\right). \tag{2.13}$$

where $\hat{\Sigma}_i$ is the error covariance for the VAR model with p_i lags.[5] This test is distributed χ^2 with $m^2(p_1 - p_0)$ degrees of freedom. The degrees of freedom can be derived by considering the impact of removing the number of lags $p_1 - p_0$ from each variable for each equation. Because there are m variables, the restricted model under the hypothesis test reduced the number of regressors in each equation by $m(p_1 - p_0)$. Because this occurs in each of the m equations in the VAR, the total number of restrictions in the model under the null hypothesis is $m^2(p_1 - p_0)$ fewer lags.

Often small-sample corrections are applied to this test. First, the test may be presented as an F test. This, however, is a rescaling of the χ^2 statistic by its degrees of freedom, so it is asymptotically identical to the χ^2 test. Second, as in the estimation of the standard error of the estimate in OLS regression or the small-sample correction for bias in the computation of a sample variance, one can correct the likelihood ratio χ^2 test for the number of parameters estimated in each equation in the VAR. Note that we have presented this correction in the test above. Here, the number of parameters in each equation of the unrestricted model $(mp_1 + 1)$ in the likelihood ratio test statistic in Equation 2.13 is subtracted from T, which is the number of observations. This reduces the magnitude of the χ^2 value to account for the number of parameters estimated in the model. This adjusted statistic is still distributed χ^2 with $m^2(p_1 - p_0)$ degrees of freedom. This correction was suggested by Sims (1980).

Hypothesis testing for lag length, however, is often complicated by one fact: The more parameters (i.e., lags) added to a time series model, the better its fit and higher the final log-likelihood value. This is the same idea that adding regressors (even spurious ones) to a linear regression model improves the R^2 for the model. As such, the likelihood ratio tests tend to choose incorrect lag lengths because when the likelihood is increasing as more lags are added, the likelihood ratio does not capture the cost of more lags. The tests need to be further corrected for pretest biases when used for sequential testing of lag length (Lutkepohl, 1985, 2005). This pretest bias arises because one is estimating a sequence of tests, so the inferences for the pth versus $p + 1$st test depend on the results of the $p - 1$ versus p test. The result is that one should use a larger critical value or rejection probability for higher-order tests.[6] This is interpreted as requiring more evidence to add another lag to the VAR.

A second type of measure is often used to determine the lag length in VAR models. These types of measures are *information criteria* such as Akaike's information criterion (AIC), the Bayesian (or Schwarz) information criterion (BIC or SC), or the Hannan-Quinn criterion (HQ). Information criteria measures are an effort to determine the trade-off between model fit and parsimony. They are based on the likelihood function for a model, penalized by the number of parameters. For two models that fit the data equally well (i.e., they have the same likelihood value), the more parsimonious model pays a smaller penalty and is thus superior based on an information criteria measure.

The AIC, BIC, and HQ measures differ in the penalty that they apply for additional parameters. The following are used to compute the AIC, BIC, and HQ measures for unrestricted VARs of lag length $p = 0, \ldots, p_{\max}$:

$$\text{AIC}(p) = T \log|\hat{\Sigma}| + 2(m^2 p + m), \tag{2.14}$$

$$\text{BIC}(p) = T \log|\hat{\Sigma}| + \log(T)(m^2 p + m), \tag{2.15}$$

$$\text{HQ}(p) = T \log|\hat{\Sigma}| + 2(\log(\log(T)))(m^2 p + m), \tag{2.16}$$

where T is the sample size under a model with $p_{\max}$ lags, $\log|\hat{\Sigma}|$ is the log determinant of the error covariance for a model with p lags from Equation 2.11, and m is the number of endogenous variables in the VAR. These model-fit summaries are considered "better" for smaller values, because the first term in Equations 2.14 to 2.16 is the log determinant of the error covariances for the likelihood function in Equation 2.12. The final term is a *penalty* for the number of parameters in the model: The more parameters in the model, the greater the penalty to the fit of the criterion statistics. The choice of p that yields an information criterion with the smallest value—as a function of the number of lags $p < p_{\max}$ included in the model—is considered the best model. Note that each additional lag added to the AIC or the BIC automatically raises the penalty to the fit criterion by a factor of m^2. Changes for additional lags that are smaller than these factors mean that one should choose the model with fewer lags, because the difference in the fit is not large enough to overcome the cost of adding more parameters to the model.[7]

Often a mixture of formal likelihood ratio tests and information criteria will be used to determine the lag length. Because the OLS-based estimation of the VAR is asymptotically consistent—it converges to the correct estimates as the sample size grows—the only cost of including additional lags in the model is inefficiency. Yet including too few lags of the past values may leave some dynamics unmodeled. For a formal comparison and review of the different criteria for selecting lag lengths, see Lutkepohl (2005). It should be noted that the information criteria may not select the same lag lengths, because (1) the computation depends on the choice of p_{max}, which should be sufficiently high (e.g., at least 1 year for monthly data), and (2) the fit criteria may overestimate the true VAR lag order (see Lutkepohl, 2005, chap. 4).

Two final points about lag length testing are necessary. First, all the lag length tests must be computed over the same observations. Because "smaller" models use fewer observations for the lags, there is a temptation to use all the possible observations. This is not correct, and all the χ^2 and fit criteria should be computed over identical samples when doing a search for lag length. Second, one should select an upper bound for lag lengths to consider (p_{max}) and consider a set of models from one lag to this upper bound. The lag length selected should be the one that is the most parsimonious model that can account for the dynamics.

2.5.3 Testing Serial Correlation in the Residuals

VAR estimation is quite robust as long as the residuals are uncorrelated over time. Thus, one should check the residuals for the series in a VAR model to ensure that they are uncorrelated over lagged time periods. Note that in VAR models, by design, we expect residuals to be correlated across variables via contemporaneous effects in the Σ or A_0 matrices. What we are concerned with here is that the residuals are not serially correlated.

Several methods exist for doing this, and all parallel the approaches typically used to evaluate the residuals of a univariate ARIMA model to determine whether the dynamics are correctly specified. Four main methods are generally employed, with varying degrees of sophistication. The first method includes graphical summaries of the residuals plotted over time for each variable. The second involves plots of the autocorrelation and cross-correlations of the variables over different time lags. The third method involves using *Portmanteau statistics*. The final method is to evaluate a sequence of lag tests, as described in Section 2.5.2.

To assess the possible serial correlation among the residuals, one can first use graphical methods—plotting the residuals over time. The problem with this approach is that it depends on the analyst's ability to recognize the serial correlation, which may be hard in some cases where the data are noisy, are of high frequency, or the serial correlation pattern is across multiple variables at the same time.

The second approach seeks to improve on this simple graphical method. This approach uses sample autocorrelation and cross-correlation functions for each of the series to see if there is any additional residual autocorrelation. Autocorrelation functions look at the correlation of a variable (in this case, its residuals) with its own past values, whereas cross-correlation functions look at the correlation of a variable's residuals with the past residuals of the other variables. These correlations can be graphed as just standard autocorrelation functions. Hypothesis testing using these plots can also be done.

The third method, a more formal one, which can be used to determine whether or not the residuals are white noise, is the *Portmanteau test*. Portmanteau tests such as the Box-Ljung or Q-statistic are common in univariate time series modeling, and various versions have been proposed for multivariate time series models. For all cases, these tests involve computing the correlations of the residuals from a fitted VAR model over a given lag length. For each of these tests, the null hypothesis is that the estimated residuals are uncorrelated over h lags or that the $T \times m$ matrix of residuals from Equation 2.7 are serially uncorrelated. Mathematically, this null hypothesis and its alternative are stated as

$$H_0: \quad E[e'_t, e_{t-i}] = 0, \qquad\qquad i = 1, \ldots, h > p,$$
$$H_A: \quad E[e'_t, e_{t-i}] \neq 0, \quad \text{for some } i = 1, \ldots, h > p.$$

The expression $E[e'_t, e_{t-i}]$ is the sample covariance between the residuals at time t and those at lag $t - i$. To test this hypothesis, we can construct a multivariate Q-statistic for lag h:

$$Q_h = T \sum_{j=1}^{T} \text{tr}(\hat{\Gamma}'_j \, \hat{\Gamma}_0^{-1\prime} \, \hat{\Gamma}_j \, \hat{\Gamma}_0^{-1}),$$

where $\text{tr}()$ is the matrix trace operation,[8] Γ_j is the covariance matrix of residuals at time t with those at time $t - j$, and h is the lag length at which the serial correlation in the residuals is to be evaluated under the null. The sample covariances of the residuals, $\hat{\Gamma}_i$, are computed using

$$\hat{\Gamma}_i = T^{-1} \sum_{t=i+1}^{T} \hat{e}'_t \hat{e}_{t-i},$$

where e_t are the residuals from the VAR model. Note that the lag "0" covariance is the same as the residual covariance matrix; thus, $\Gamma_0 = T^{-1} \sum_{t=1}^{T} \hat{e}_t' \hat{e}_t = \hat{\Sigma}$. The test statistic Q_h is distributed asymptotically as a χ^2 distribution with $m^2(h - p)$ degrees of freedom.

As is the case with the univariate version of this test, the Q-statistic may perform poorly in small samples. An alternative version with better small-sample properties can be constructed. This modified Q or Box-Ljung-style Portmanteau statistic accounts for the estimates of the serial correlation of the residuals at the interim lags or from 1 to h. This modified Q-statistic test has the form

$$Q_h^* = T^2 \sum_{j=1}^{T} \frac{1}{T - j} \text{tr}(\hat{\Gamma}_j' \hat{\Gamma}_0^{-1'} \hat{\Gamma}_j \hat{\Gamma}_0^{-1}),$$

and is asymptotically χ^2 with $m^2(h - p)$ degrees of freedom. Here, the degrees of freedom correction, a factor of $T/(T - j)$, accounts for the estimates of the correlations from lags 1 to j.

For both these tests, the choice of h is critical, because if there is no serial correlation at lag h, then we would fail to reject the null. But it is possible that there is serial correlation at some lag greater than the initial h. Thus, one should select a series of values for h that correspond to the periodicity of the data (e.g., 6, 12, 18, and 24 for monthly data).

An alternative test can be constructed using the ideas of the Breusch-Godfrey Lagrangian multiplier (LM) test for serial correlation in a univariate regression (Lutkepohl, 2004). In the univariate version of this test, one regresses the residuals from an OLS model on the lagged values of the dependent variable and the lagged values of the residuals and tests whether the regression coefficients for the lagged residuals in the unrestricted model are zero. In the multivariate version, one fits two additional VARs using the residuals from a VAR(p) model of y_t.

There are four steps in the LM test for serially uncorrelated VAR residuals. The first step is the estimation of an *unrestricted artificial VAR* where we allow for the possibility that the residuals from the original VAR of y_t are correlated. This is done by estimating the VAR

$$e_t = y_{t-1}A_1 + \cdots + y_{t-p}A_p + e_{t-1}B_1 + \cdots + e_{t-h}B_h + u_t, \quad (2.17)$$

which is a VAR formed by regressing the matrix of the residuals on lags 1 to p of y and the lags 1 to h of the residuals from the original model.

The second step is estimating a *second artificial VAR*, which is the restricted model where $B_1 = \cdots = B_h = 0$:

$$e_t = y_{t-1}A_1 + \cdots + y_{t-p}A_p + u_t^R. \quad (2.18)$$

This restricted model corresponds to the null hypothesis that the residuals of the original VAR are uncorrelated.

Third, one constructs the residual covariances for these two artificial VAR residual models (Equations 2.17 and 2.18)

$$\widetilde{\Sigma}_e = T^{-1} \sum_{t=1}^{T} \hat{u}'_t \hat{u}_t,$$

$$\widetilde{\Sigma}_R = T^{-1} \sum_{t=1}^{T} \hat{u}^{R'}_t \hat{u}^{R}_t,$$

where the first residual covariance matrix is estimated from the unrestricted artificial regression and the second residual covariance matrix is estimated from the second, restricted artificial regression.

Finally, the χ^2 LM test statistic to evaluate the presence of serial correlation in the VAR residuals can be computed by

$$\text{LM} = T[m - \text{tr}(\widetilde{\Sigma}_e \widetilde{\Sigma}_R^{-1})],$$

where m is the number of (endogenous) variables in the system and $\text{tr}()$ is the trace operation. The test statistic is distributed χ^2 with hm^2, the number of restrictions on the parameters of the restricted model under the null hypothesis of no residual serial correlation.

A final and probably the most typical method of ensuring that a VAR has white noise residuals is overfitting the VAR. Here, one fits a series of VAR models and tests the restriction that a model with a restriction to p^* lags is the same as a model with $p^* + 1$ lags. This is the same test procedure described earlier for the lag length specification analysis. This testing down to a parsimonious model typically ensures that the residuals will be white noise.

In summary, all the tests proposed for the serial correlations of the residuals are asymptotically equivalent and the conclusions hold only as the sample increases in size. For a detailed discussion of these tests, see Lutkepohl (2004, pp. 127–131).

If any of these methods for evaluating the residuals show evidence of serial correlation, then additional lags will need to be added to the VAR. These lags should be added to each equation, not just the variable(s) still exhibiting serial correlation. Recall that in the VAR approach, one typically eschews using tests to (overly) restrict the VAR model. As such, the most common of these approaches is the overfitting of the VAR model and testing the number of appropriate lags. This is the same technique outlined earlier to determine the proper lag length for the model. In most cases, the reported VAR results will either discuss the lag length selection testing or show results for different lag lengths to indicate the robustness of the dynamic specification.

2.5.4 Granger Causality

An important question in assessing the relationships among multiple time series in an unrestricted VAR model is the value of individual variables for explaining the other variables in the system of equations. This question can be posed in one of three equivalent ways:

- What value does a variable Y_t have for *predicting* another variable Z_t in a dynamic system of equations?
- Is a variable Y_t exogenous in a time series model with respect to Z_t?
- Is the variable Y_t informative (linearly) about future values of Z_t?

These three questions are considered identical in assessing the relationships among variables in a multivariate time series model. The equivalence of the answers to these questions is based on the determination of *Granger causality* in a time series model (see Granger, 1969; Sims, 1972).

To define the concept precisely, consider the following bivariate VAR model for two variables (Y_t, Z_t):

$$Y_t = \alpha_0 + \sum_{i=1}^{p} \alpha_i Y_{t-i} + \sum_{i=1}^{p} \beta_i Z_{t-i} + \epsilon_{1t}, \qquad (2.19)$$

$$Z_t = \beta_0 + \sum_{i=1}^{p} \gamma_i Y_{t-i} + \sum_{i=1}^{p} \delta_i Z_{t-i} + \epsilon_{2t}. \qquad (2.20)$$

Using this system of equations, Granger causality is defined as follows:

Granger causality For linear models, Y_t Granger causes Z_t if the behavior of past Y_t can better predict the behavior of Z_t than Z_t's past alone.

The reverse is true as well. For the model in Equations 2.19 and 2.20, if Z_t Granger causes Y_t, then the coefficients for the past values of Z_t in the Y_t equation are nonzero, or $\beta_i \neq 0$ for $i = 1, 2, \ldots, p$. Similarly, if Y_t Granger causes Z_t in the Z_t equation, then the coefficients for the past values of Y_t are nonzero, or $\gamma_i \neq 0$ for $i = 1, 2, \ldots, p$.

One tests for Granger causality by assessing whether the past values of a variable, $Y_{t-1}, \ldots, Y_{t-p}$, predict the present values of a variable Z_t in a VAR. The formal testing for Granger causality is then done by using an F test or a χ^2 test for the joint hypothesis that the possible causal variable *does not cause the other variable*. The null hypothesis of the test is one of *noncausality*, as defined above. We can specify the null hypothesis for the Granger causality test as follows:

H_0: **Granger noncausality** Z_t does not predict Y_t if
  $\beta_1 = \beta_2 = \cdots = \beta_p = 0$.

H_A: **Granger causality** Z_t does predict Y_t if
  $\beta_1 \neq 0, \beta_2 \neq 0, \ldots,$ or $\beta_p \neq 0$.

Note that the alternative hypothesis is that *any* of the coefficients may be nonzero. This hypothesis test can be implemented by an F test or a likelihood ratio test. The F test implementation is simple and requires running two regression models[9]:

Model 1 (unrestricted)

$$Y_t = \alpha_0 + \sum_{i=1}^{p} \alpha_i Y_{t-i} + \sum_{i=1}^{p} \beta_i Z_{t-i} + \epsilon_{1t}.$$

Model 2 (restricted)

$$Y_t = \alpha_0 + \sum_{i=1}^{p} \alpha_i Y_{t-i} + u_{1t}.$$

To compute the test, we find the residuals sums of squares (RSS)

$$\text{RSS}_{\text{Unrestricted}} = \sum_{t=1}^{T} \epsilon_{1t}^2,$$

$$\text{RSS}_{\text{Restricted}} = \sum_{t=1}^{T} u_{1t}^2.$$

The test statistic for comparing the sums of squared residuals has an F distribution with $(p, T - 2p - 1)$ degrees of freedom:

$$F(p, T - 2p - 1) \sim \frac{(\text{RSS}_{\text{Restricted}} - \text{RSS}_{\text{Unrestricted}})/p}{\text{RSS}_{\text{Unrestricted}}/(T - 2p - 1)}.$$

If this F statistic is greater than the critical value for a chosen level of significance, we reject the null hypothesis that Z_t has no effect on Y_t and conclude that Z_t Granger causes Y_t. Note that a similar test can be constructed to evaluate the Granger causality of Y_t on Z_t by testing that $\gamma_i = \gamma_2 = \cdots = \gamma_p = 0$ in Equation 2.20.

Often χ^2 tests will be used for evaluating Granger causality. Because of the large number of variables and lags in a VAR, the F test described above will lose power as the numerator and denominator degrees of freedom approach the same value. This will lead to F tests that are biased toward the null hypothesis of noncausality. The χ^2 version of the Granger causality test can be constructed using a likelihood ratio test or a Wald test, where the restricted coefficients under the null are the same as those constrained to be zero in the F test formulation.

Both versions of the test are asymptotically equivalent, but the F test formulation is easy to implement and typically works well for testing the two-variable causality hypotheses.

2.5.5 Interpreting Granger Causality

One criticism of the idea of Granger causality testing is that in effect it does not imply causality in a "philosophic sense." This criticism is true (e.g., Hamilton, 1994, pp. 302–309) insofar as the standard definition of causality depends on finding a relationship that is temporally consistent (change in one variable happens before another), supported by a statistically significant correlation, and nonspurious. The first two of these criteria can be easily supported by a test for Granger noncausality, but not the third. The F and χ^2 tests discussed in Section 2.5.4 only test that the coefficients of the lagged values are nonzero—so they do not allow one to assess the *direction* of the relationship. To determine that the relationship is nonspurious, theory is necessary.

A second argument that should be considered when assessing the temporal relationships among variables using Granger causality is the role of contemporaneous correlation. Consider the following example: Suppose that Granger causality hypothesis tests show that one cannot reject the null hypothesis of noncausality. This means that the past values of other variables in the system cannot predict other variables. But in such a system, the variables may be strongly autoregressive and well predicted by themselves. If there is a high contemporaneous correlation among the series in the multiple time series model, then the innovations or the shocks will be highly correlated. Thus, although the "causality" with respect to the past values may not be present, the series may be highly contemporaneously correlated—both series may be "caused" by commonly correlated innovations. Thus, a finding of Granger noncausality does not imply that the series in the system are uncorrelated. Instead it means that their past values are not predictive of each other. For an example of how this can occur and how theory can predict noncausality, see Williams and McGinnis (1988).

A third issue in the interpretation of Granger causality tests is the role of model specification. There are two potential issues: an incorrect lag specification and the omission of a relevant Granger causal variable. Suppose first that the lag length used in the VAR is incorrect. There are two cases to consider: too many lags and too few lags. If one includes too many lags in the VAR model, the resulting estimates will possibly be inefficient, but unbiased—just as in a linear regression model. Thus, hypothesis tests will be unbiased, but inefficient. We will then be likely to fail to reject the null when we really should. This is of minor consequence, because we would then generate the null finding of noncausality.

In contrast, consider the case of too few lags. In this case, the VAR estimates will be biased and inefficient, as expected from linear regression results when there are omitted variables. The omitted lags will also leave unmodeled dynamics, which will lead to serially correlated residuals. The resulting test statistics will then be "too good" in the sense that they will lead us to reject the null hypothesis of noncausality too often. Thus, omitted lags and dynamics in the specification of a VAR model will lead to one's finding causality when it may not really be present.

Another possible misspecification problem in a VAR is the omission of an equation or variable that is a Granger causal variable and affects all the variables in the system. Suppose, for instance, that we have a two-variable system that is exogenous or has a test statistic that supports the null of Granger noncausality. Suppose one thinks it is possible that there exists a third variable that if included would overturn this noncausality finding. Such a variable is highly unlikely, though, because Litterman and Weiss (1985) show that the omission of this third variable is highly unlikely to overturn a finding of exogeneity or noncausality. The rationale is that for the inclusion of a third variable to overturn this finding of noncausality among the first two, it would have to perfectly cancel out the exogeneity result found among the first two variables over all the lags, which is highly unlikely.

A final issue that can impact Granger causality inference is the presence of unit roots or stochastic trends in the variables of a VAR. If one or more of the variables in the VAR model has a unit root, then the test statistics for the model parameters will have a nonstandard distribution (Sims et al., 1990). These tests have nonstandard distributions because the underlying problem is related to what happens when one regresses an $I(1)$ or unit root variable on a stationary variable. Granger and Newbold (1974) and Philips (1986) show that when one regresses a unit root variable on variables that are not related to it, the regression coefficients will be incorrectly statistically significant because of the trend in the dependent variable. Thus, efforts to assess Granger causality can be seriously compromised by unit root variables.

The consequence of unit roots is that from a frequentist perspective, testing Granger causality is complicated by nuisance parameters when unit roots are present. Several authors have noted that simple modifications of Granger causality tests in the presence of unit roots do provide correct test statistics (Dolado & Lutkepohl, 1996; Lutkepohl & Reimers, 1992a;

Zapata & Rambaldi, 1997). The basic idea is that the presence of unit roots will leave unmodeled dynamics. In this case, standard F tests or χ^2 tests will be incorrect. These authors propose various methods to augment the hypothesis tests for Granger causality based on the number of unit roots in the model. If there are m variables in the model, then there can be at most $d \leq m$ unit roots. The correction for this test is to estimate a model with $p + d$ lags and use a Granger causality test based on the first p lags of the VAR model. Another approach is to conduct a Granger noncausality test using the first differences of the data, which are stationary and not subject to the nonstandard test results described above.

Despite these caveats, Granger causality is a useful concept for theory building and testing. The key point to remember is that Granger causality aids only in the prediction and forecasting of variables—it does not aid in the inference about structural parameters in a multiple time series model. Because the VAR model is estimated by OLS, the estimates are consistent even in the face of true Granger noncausality restrictions. Further, so long as the lag length is sufficient to ensure that the errors are white noise, the tests will be valid. This is quite important—empirical applications of Granger noncausality tests should then check whether Granger causality tests are sensitive to lag length.

2.5.6 Testing Other Restrictions in a VAR Model

Because the VAR models are built using standard results from multivariate regression models, hypothesis testing (except for some cases with nonstationary variables) can be done using χ^2 and F tests from standard regression analysis. This includes hypothesis tests for the significance of individual or groups of coefficients.

One particular kind of hypothesis test in VAR models is a block exogeneity restriction. These can involve more than just a restriction on a particular set of lags (as in the tests for lag length) or the effect of one variable's lags in a single equation. Likelihood ratio tests can be constructed that look at the effects of a set or block of coefficients for different variables across multiple equations. These can be evaluated using a χ^2 test constructed to restrict the coefficients across a set of equations to zero. The basic idea of these tests is that if the restrictions are true, then the likelihood ratio of the restricted and the unrestricted models should be small. If the null hypothesis restriction on the coefficients of the VAR is false, then this likelihood ratio will be large and we will reject the null.

2.5.7 Impulse Response and Moving Average Response Analysis

All the earlier tests are helpful in understanding the relationships among the variables in a multiequation time series system. But our real goal in employing a VAR model is to describe and characterize the dynamic behavior of the series. The most useful way to describe the dynamics of a VAR system is through MARs or IRFs. MARs are the equivalent of dynamic multiplier analyses in other time series models (e.g., ARIMA). Once one has estimates for a VAR model, the residuals or shocks to each of the equations are random noise. Thus, a residual is an exogenous shock or surprise innovation to the equations in the system. MARs are used to interpret the impacts of these shocks to the vector moving average (VMA) representation of the model. These responses are standardized and presented in terms of the scales of the variables in the fitted VAR.

The VMA representation of a VAR is based on its infinite moving average representation. Every stationary (vector) finite-order autoregressive time series model can be written as an infinite lagged moving average time series model (Wold, 1954). This is useful because one can then recenter the model around its equilibrium values and look at the impacts of shocks or surprises on the series. The dynamics of these surprises depend on the dynamics of the VMA representation of the VAR.

The VMA representation of the VAR can be found by rewriting the VAR as a moving average process. This is done by factoring the lag polynomial for the autoregressive process. A lag polynomial is a matrix function that creates the lags for a set of time series. For example, we can rewrite the two lags for a VAR(2) model using the following polynomial:

$$y_t = c + y_{t-1}B_1 + y_{t-2}B_2 + e_t,$$
$$y_t = c + y_t(B_1L + B_2L^2) + e_t,$$

where L^k is the lag operator that shifts any variable it is multiplied by back k periods: $L^k y_t = y_{t-k}$. This allows us to have a compact notation for the creation of the lag polynomial for the system of equations. We then see that the errors depend on how the feedback occurs in the other variables through the autoregressive coefficient matrices.

Using this notation, we can rewrite a VAR(p) model as a VMA:

$$y_t = c + y_{t-1}B_1 + y_{t-2}B_2 + \cdots + y_{t-p}B_p + e_t,$$
$$y_t - y_{t-1}B_1 - y_{t-2}B_2 - \cdots - y_{t-p}B_p = c + e_t,$$
$$y_t(I - B_1L - B_2L^2 - \cdots - B_pL^p) = c + e_t,$$
$$y_t = (c + e_t)(I - B_1L - B_2L^2 - \cdots - B_pL^p)^{-1},$$
$$y_t - d = e_t(I + C_1L + C_2L^2 + \cdots),$$

(2.21)

where d is the VAR constant term, c, divided by the autoregressive lag polynomial. The last step of this derivation produces a set of coefficients for an infinite-lag VMA representation that corresponds to the finite-order VAR(p) model. This is done by equating the polynomial of the autoregressive coefficients to a corresponding set of moving average coefficients:

$$(I - B_1 L - B_2 L^2 - \cdots - B_p L^p)^{-1} = (I + C_1 L + C_2 L^2 + \cdots),$$
$$(I - B_1 L - B_2 L^2 - \cdots - B_p L^p)(I + C_1 L + C_2 L^2 + \cdots) = I,$$

where the left-hand side is the autoregressive representation of the m-dimensional matrix polynomial of order p and the right-hand side is the m-dimensional moving average polynomial of infinite order.

The MAR coefficients C_i in Equation 2.21 can be computed from the autoregressive coefficients via the following recursions:

$$C_1 = B_1$$
$$C_2 = B_1 C_1 + B_2$$
$$C_3 = B_1 C_2 + B_2 C_1 + B_3 \tag{2.22}$$
$$\vdots$$
$$C_\ell = B_1 C_{\ell-1} + B_2 C_{\ell-2} + \cdots + B_p C_{\ell-p}$$

with $C_0 = I$ and $B_j = 0$ for $j > p$. This infinite-lag moving average representation of the model shows the same dynamics as the autoregressive representation.

The reason for inverting the model from an autoregressive representation to a moving average representation is that it simplifies the dynamic structure and allows us to track out the impact of the exogenous shocks or innovations. With this VMA representation, one can analyze the impact of an exogenous shock to e_t on the value of y_t for each equation. We do this with the VMA representation rather than the VAR representation so that we can see how the shocks that differ from the mean value of zero evolve and die out over time (as opposed to the VAR model, which would show the variation in the responses to the shocks around the equilibrium level of each series). This can be seen by looking at the left- and right-hand sides of the VMA representation in Equation 2.21. The right-hand side is the equilibrium level of the VAR or the long-run level of each series. The left-hand side then describes how the deviations of the series evolve around this equilibrium level.

Another way to interpret the VMA or moving average representation is as the impact of a change in the innovation in one equation. This is a matrix derivative for an innovation of a given size. The change in the equation for the endogenous variable y at time ℓ for a change in the innovation in variable j at time s, denoted $e_j(s)$, is the derivative of y_ℓ in Equation 2.21 with respect to this shock or

$$\frac{\partial y_i(\ell + s)}{\partial e_j(s)} = c_{ij}(\ell), \qquad (2.23)$$

where $c_{ij}(\ell)$ represents the element of the ith row, jth column of the C_ℓ matrices defined in Equations 2.21 and 2.22. This quantity measures the magnitude of a response in equation i for a shock to variable j at time $s > \ell$. Note that this is a vector process, so there is one impulse response for a shock to each variable in the system. Thus, if there are m variables in the system, then there are m^2 impulse responses (including the response of each variable to its own innovations or surprises).

This derivation of the MARs for a VAR provides three different ways of interpreting the dynamic relationships summarized in VAR coefficients. The first is the explicit moving average representation where one analyzes the effects of shocks to the system. The second is that the MAR acts as a way to interpret the dynamic multiplier effects of changes to various variables in the VAR. This interpretation is given support by the equivalence of the finite VAR(p) model by the VMA. The final way of thinking about these derivations is as a marginal effect of changing one of the variables in the model— the analysis of an exogenous shock to one equation as in Equation 2.23. Here, one is analyzing the marginal changes in each of the variables in the system of equations for an e_t shock to the equations. These changes in the endogenous variables are dynamic and thus should be traced out over time.

One issue in all these interpretations of the MARs is the identification of the shocks or innovations used in the system. In the derivation of Equation 2.22, we assumed that the contemporaneous responses of the variables were uncorrelated or that $C_0 = I$. In general, the contemporaneous shocks are not uncorrelated, as shown by the off-diagonal elements of the estimated covariance matrix Σ, which is the $m \times m$ covariance matrix of the VAR residuals. Thus, an assumption needs to be made about how the error process is correlated or related.

Sometimes this decision can be made based on theory—there may be reasons to assume that innovations in one series are not correlated with another. It may be the case that there is no a priori restriction that should be imposed on the innovation process. Further, the ordering of variables in the VAR and taking the decomposition of the VMA matters. In a sense, then, it appears that interpreting the dynamics of VAR models presents the same difficulties that are present in SEQs. However, in VAR modeling, the decisions about the identification of contemporaneous relationships are made *after accounting for the dynamics in the data*. This is critical because it means that identification decisions do not bias or restrict our interpretation of the dynamics. Rather, they are explicit assumptions about the contemporaneous relationships that can then be evaluated within a fully specified dynamic model.

The role of this identification decision can be seen in the impulse response matrices. The $m \times m$ matrix C_ℓ, with elements c_{ij} at time ℓ, that contains the responses of equation i (the rows) to a shock in variable j (the columns) at time ℓ is found by

$$C_0 = A_0^{-1}, \quad C_\ell = \sum_{j=1}^{\ell} C_{\ell-j} B_j, \quad i = 1, 2, \ldots, \quad (2.24)$$

where we use the convention that $B_j = 0$ for $j > p$. Here, the responses to the shocks in period 0 are those provided by the contemporaneous correlations of the innovations in A_0^{-1}. These correlations are then filtered through the autoregressive coefficients by the second recursive calculation for the next s periods. The result is a series of s matrices for each of the periods for which the moving averages are computed. Thus, there is an initial and explicit dependence on the initial identification of the contemporaneous residuals.

The contemporaneous correlations can be computed in several different ways based on the data and theory:

1. In the absence of any knowledge beyond the ordering of the variables' contemporaneous shocks in the system, one can compute A_0^{-1} from the Cholesky decomposition of the error covariance matrix $\hat{\Sigma}$. In this case, we find the matrix decomposition $A_0^{-1'} A_0 = \hat{\Sigma}$. Here, we are finding the square root of the matrix $\hat{\Sigma}$, which is lower triangular (in $A_0^{-1'}$). This is also known as orthogonalization of the residuals.
2. A second method would be to factor the matrix $\hat{\Sigma}$ into some other decomposition through an estimation of A_0^{-1}. This is often difficult to estimate and identify in the absence of strong theory.
3. A final method is to find a matrix A_0 that solves the system of equations exactly.

The most common method and the one we will generally illustrate is based on a Cholesky decomposition of $\hat{\Sigma}$. In this case, the ordering of the variables will be crucial. If the correlations among the residuals are low, then the ordering will not be a major factor in the computation of the responses. However, for series that are highly correlated, these orderings will affect the interpretation of the results. As a result, one will want to evaluate multiple orderings of the variables (based on theory) and to check robustness.

It is important to note that regardless of the method chosen, there must be some decision made about the identification of the contemporaneous shocks in the VAR. If the innovations in the variables in the system are uncorrelated, then the choices about the contemporaneous correlations will be rather innocuous. If the contemporaneous responses are highly correlated, then several approaches can and should be considered. First, if only a subset of the m variables in the VAR are highly contemporaneously correlated, then these should be placed together in the Cholesky ordering. Although not much can be said about the impacts of these variables on each other, more can be determined about their impact on the other variables in the system. Second, one can also perform multiple analyses and see how the dynamics of the system's impulse responses are affected by the choices about the contemporaneous orderings. Finally, rather than using a Cholesky ordering, one can consider other decompositions of the responses.

The impulse responses computed after identifying the residual correlations or contemporaneous relationships are a set of s matrices that are each $m \times m$. This means that summarizing and presenting them presents a unique challenge. The standard way to do this is to present the responses graphically as a set of $m \times m$ time series of responses of length s by rearranging the elements of C into m^2 time series of length s. These m^2 series are then arranged into a set of $m \times m$ graphs. The graphs are arranged to reflect the identification assumption in the A_0^{-1} matrix—that is, using the same ordering of the variables in the rows of A_0. Examples of these impulse responses and their presentation are presented in the next chapter.

2.5.8   Error Bands for Impulse Responses

Impulse responses are used to trace out the dynamic responses of the equations in the VAR to a set of identified shocks. Because these are moving averages for generally stationary autoregressive processes, we expect that these shocks will die off or return to zero. In addition, the identification of the shocks typically assumes that the magnitude of the shocks is 1 standard deviation of the residuals in the VAR. These initial responses are then traced out as functions of time.

The discussion until this point has been focused on estimating and computing the mean of the responses. We would also like to know whether the MARs for equation i for a shock to variable j differ significantly from zero. Responses that have confidence intervals that differ from zero will be considered statistically significant—or different from zero along their time paths from $0, \ldots, s$. Responses whose confidence regions include zero for the time horizon of the responses are then not statistically different from zero.[10]

Because the impulse responses for a VAR model are based on the estimated autoregressive coefficients and their moving average representations, they are themselves random quantities. The responses, which show the dynamic response of one variable to a shock or innovation in another, thus have a mean and a variance. The mean estimate is provided by the earlier impulse response computation that finds the effect of a series of shocks for the system of equations as a function of the estimated autoregressive parameters. To estimate the confidence region of impulse responses, one must determine how the two sources of variation in autoregressive coefficients—the coefficients themselves and the residual covariance matrix—lead to variation in the responses.

Deriving and estimating the variances of the responses is more difficult. The difficulties are on two fronts. First, the analytical derivation of the variance of the impulse responses is complicated and may be based on (possibly incorrect) assumptions about the actual data. Because the maximum-likelihood estimates of the VAR parameters (via equation-by-equation OLS) are only asymptotically normal, it means that in a finite sample, the VAR coefficients may be only approximately normal (Lutkepohl, 1990). But the impulse response coefficients are nonlinear functions of only these approximately normal parameters (matrix polynomials, actually). Thus, we have no guarantee in most empirical applications that the responses are normally distributed and can be easily summarized by their mean and variance alone.

Second, the responses to the shocks are themselves likely to be serially correlated. The reason why is implicit in the structure of the VAR model itself. The model is based on specifying how the past lags or the history of the variables in the system explain the current values today. If these lagged values consist of some combination of deterministic components and random shocks or innovations, then subsequent innovations in the variables depend on these earlier innovations, as is implied by the VMA representation of the vector autoregression. Although in the course of estimation we assume that the innovations or residuals are uncorrelated, in the dynamic analysis these responses are serially correlated by design. This means that attempts to summarize the variance must also account for the implicit serial correlation in the responses.

Several methods have been proposed for developing measures of uncertainty or error bands and describe the confidence region of the responses in Equation 2.23 (Runkle, 1987). Analytic derivatives and related normal asymptotic expansions for computing the variances of the responses are presented in Lutkepohl (1990) and Mittnik and Zadrozny (1993). The approximations of these analytical derivative methods tend to perform poorly as the impulse response horizon, s, increases. Kilian (1998) presents a small-sample "bootstrap after bootstrap"-based confidence interval for impulse responses. This bootstrap method reduces the bias in the initial estimates of the VAR coefficients, but it does not adequately account for the potential non-Gaussian, nonlinear, highly correlated aspects of the responses (see the discussion in Sims & Zha, 1999, pp. 1125–1127).

The standard approaches to computing the error bands are based on constructing the following interval:

$$\hat{c}_{ij}(t) \pm \delta_{ij}(t), \qquad (2.25)$$

where $\delta_{ij}(t)$ are some functions that define the upper and lower bounds of the confidence intervals for some $100(1 - \alpha)$ region for the response at time t, $c_{ij}(t)$. These are presented graphically by plotting the three functions $\hat{c}_{ij}(t) - \delta_{ij}(t)$, $\hat{c}_{ij}$, and $\hat{c}_{ij}(t) + \delta_{ij}(t)$ as functions of t. These are effectively known as "connect the dots" error bands and are a standard output from common statistical software (e.g., RATS, Eviews, Stata). Measures of the impulse response variation computed in this manner typically assume that the responses are uncorrelated over time—the function $\delta_{ij}(t)$ is uncorrelated with $\delta_{ij}(t + k)$. This means that the variation of future responses does not depend on the past variance of the responses except through the variation in the autoregressive parameters.

There are several ways to compute the error bands and the functions $\delta_{ij}(t)$. The most common approach is to use a simulation to compute a sample of the responses and summarize this sample. This is done by computing a Monte Carlo sample from the (posterior) distribution of the VAR coefficients.[11] In this method, a (posterior) sample of the VAR parameters is constructed using the asymptotic distribution of the VAR coefficients. These coefficients are multivariate normal with mean $B \sim MVN(\hat{B}, \Sigma \otimes (X'X)^{-1})$ based on the representation in Equations 2.7 and 2.10. The inverse of the posterior estimates of the residual covariance, Σ^{-1}, has Wishart distribution with $\Sigma^{-1} \sim \text{Wishart}(S, T)$, where $\bar{S}$ is the sample estimate of $\hat{\Sigma}^{-1}$ (for details, see Zellner, 1971). Using these facts, we can use the following steps to construct a sample of the VAR impulse responses:

1. Draw Σ^{-1} from the inverse Wishart distribution.
2. Take a draw of the autoregressive coefficients from $B \sim MVN$ $(\hat{B}, \Sigma \otimes (X'X)^{-1})$ using Σ from Step 1.
3. Using the draw from Steps 1 and 2, compute a set of impulse responses using Equation 2.21.
4. Save the estimated responses from Step 3.
5. Repeat Steps 1 to 4 N times, where N will provide the appropriate precision about the responses of interest.

Summaries of the N sets of $m \times m \times s$ responses will provide an estimate of the variance of the impulse responses. The number of samples, N, is usually chosen to be between 1,000 and 5,000, depending on the level of precision needed and the scale of the data. One should typically check the robustness of using larger values for N.

From this sample of impulse responses, we can then compute a normal approximation to the average $\hat{c}_{ij}$:

$$\hat{c}_{ij}(t) \pm z_\alpha \sigma_{ij}(t),$$

where z_α are the normal probability density quantiles, and $\sigma_{ij}(t)$ would be the standard deviation of $c_{ij}(t)$, the response of variable i to shock j at time t ($z_\alpha = 1$ for 68% confidence error bands and $z_\alpha = 1.96$ for 95% confidence error bands). This assumes that the impulse responses are normal in small samples. Thus, it will produce bands that are symmetric around the mean impulse response.

Although this Gaussian approximation approach has previously been used in major time series statistical packages (e.g., RATS, Eviews, Stata), an alternative (used currently in RATS) would be to compute the empirical quantiles or the empirical percentiles of the responses from the sample for $c_{ij}(t)$ computed for each response and time point. In this case, we could estimate the posterior interval based on the highest posterior density region or pointwise quantiles, namely,

$$[c_{ij,\alpha/2}(t), c_{ij,(1-\alpha)/2}(t)],$$

where the subscripts $(1-\alpha)/2$ and $\alpha/2$ denote the bounds of the α confidence set or interval (0.05, 0.1, 0.16, etc.), computed by taking the empirical quantiles. Details of this and related methods can be found in Sims and Zha (1999) and Brandt and Freeman (2006).

Interpreting the MARs and their error bands requires caution. Recall that in the VAR, we only estimated a finite number of parameters. Once these parameters are used to generate a high-dimensional set of impulse responses that have $s > p$ response periods, the projections of the innovations may be quite imprecise—for precisely the same reason that forecasts are imprecise. The width of the confidence region or error bands of the responses will generally grow exponentially over the response horizon. This is because the uncertainty of the parameters and the forecast errors of the VAR will grow over longer impulse response horizons. As such, we must exercise caution in interpreting the responses in later periods. This is because the confidence intervals for the responses will begin to grow exponentially over the time horizon.

A second issue in the interpretation of the impulse responses is the assessment of model stability. The responses in the MARs should not explode. Stable systems will have responses that decay to zero. Furthermore, one should be suspect of impulse responses that explode forever— yet in small samples, we may see this behavior. Small-sample (unrestricted) VARs will show some instability. But such unstable responses should be discounted and their error bands are likely to be too large.

With these points in mind, we propose several rules of thumb for interpreting MARs and their error bands:

1. Remember that the responses are from a reduced form system. Thus, after the first period, part of what the responses summarize is the *feedback* in the system of equations. Thus, the responses show both direct and indirect effects of the shocks. The assessment of direct effects should then be augmented by the careful use of F tests for Granger causality.

2. After the initial periods of the responses in the MAR projections, the standard errors around the responses grow relatively quickly. Thus, part of the error bands is reflecting the uncertainty about the innovations themselves (in addition to the uncertainty about the dynamics in the system).

3. Decide how likely the nonzero responses should be. Sims (1987) argues that the VAR models provide a good assessment of the uncertainty about the dynamics, but that the results are asymptotic. Thus, classical, small-sample hypothesis interpretations of the results are likely to be nonrobust. In this regard, the Bayesian-shape error bands described in Brandt and Freeman (2006) should probably be employed.

2.5.9 Innovation Accounting or Decomposition of Forecast Error Variance

A second method for interpreting the interrelated dynamic changes in multiple time models is known as innovation accounting or decomposition of the forecast error variance. This method estimates the amount of variation in each of the endogenous variables in the system of equations due to the changes in each of the endogenous variables over some period of time.

There are two ways to interpret or understand innovation accounting. The first is that one is analyzing deviations from the forecasted paths of the variables. Suppose that the variables in the multiple time series model of interest followed the forecasted path computed from the estimated coefficients. These forecasts contain two components in a VAR model: the predicted paths of the variables and the unexpected innovations or shocks to each equation. For each equation in the VAR system, one can compute how much of the variance in the predicted path of the dependent variable is from its own past and how much is due to the past values of the other variables in each equation. This partitioning of the variance of the dependent variable in each equation in the VAR—the variance from the dependent variable (and its past values) and the variance from the other variables (and their past values)—is an accounting or decomposition of the total variation in the variable over time.

A second way to think about innovation accounting is as a generalization of an analysis of variance (ANOVA). Because we are dealing with multiple time series, the ANOVA of each of the endogenous variables in the system should account for the amount of variance in each variable that can be explained by the others over some period of time. Innovation accounting is then analogous to a series of ANOVA calculations over some time horizon.

Innovation accounting for the variances for the variables in a VAR is done by decomposing the variance of the variables in each equation into the amount of variation from each of the other variables in the system. This is the same as asking how forecast errors or innovations in each variable affect the system of equations and the variance of each equation. Here, the innovations are the unexpected or unforecasted components of the series. Thus, we want to analyze the origins of the forecast errors in the model.

Innovation accounting is accomplished by computing the variance of the forecast errors for a VAR. The VMA representation of the VAR in Equation 2.21 is used for this computation. Starting with this equation, one can compute the forecast errors for the VAR system at period s:

$$y_{t+s} - \hat{y}_{t+s} = e_{t+s} + C_1 e_{t+s-1} \\ + C_2 e_{t+s-2} + \cdots + C_{s-1} e_{t+1}. \tag{2.26}$$

The left-hand side of this equation is the difference between the observed value of the vector of endogenous variables at time $t + s$ and the predicted values from the VAR. The right-hand side is the VMA representation of the forecast errors over the current period $T = s$ back to period $s - 1$. This shows how the current forecast errors or innovations in the VAR model are functions of the past innovations. The coefficients C_ℓ are the moving average coefficients defined for the VMA representation of the VAR. This equation then shows how the innovations (shocks or residuals) in the VAR model are a function of their own past value in the VMA representation.

The variance of the forecast errors in Equation 2.26 is

$$
\begin{aligned}
V(y_{t+s} - \hat{y}_{t+s}) &= E[(y_{t+s} - \hat{y}_{t+s})'(y_{t+s} - \hat{y}_{t+s})] \\
&= \Sigma + C_1 \Sigma C_1' + C_2 \Sigma C_2' + \cdots + C_{s-1} \Sigma C_{s-1}',
\end{aligned} \tag{2.27}
$$

where $\Sigma = E[e_t' e_t]$ is the covariance of the forecast errors in period t.

The decomposition of the forecast errors over the previous periods as in Equation 2.27 can be thought of as a partitioning of the relative influence of variables in the system. This is an *innovation accounting* that describes the relative importance of cross-variable effects in the system. Although this calculation will tell us how much of the variance in each variable depends on its own past values, we would also like to be able to separate out the impact of the innovations of the past values of the jth variable from the impact of the other variables. This is similar to an ANOVA: Here, we wish to determine how much of the variance in one variable is due to its own innovations and how much is due to innovations to other variables over time.

The typical way that this ANOVA of the forecast errors is accomplished is via the orthogonalization of Equation 2.27. As discussed in the section on impulse responses, orthogonalization serves two purposes. First, it standardizes the variances of the shocks or innovations. Second, it establishes the contemporaneous relationships among the forecast innovations—how these contemporaneous errors are correlated. This latter point is critical because it identifies which linear combinations of the forecast innovations are related to each other. Note that this is a situation where the identification assumptions about the contemporaneous relationships among the variables is made in VAR analysis.

The orthogonalized forecast innovations can be written as

$$
e_t = u_t A_0^{-1} = u_{1t} a_1 + u_{2t} a_2 + \cdots + u_{mt} a_m, \tag{2.28}
$$

where a_i is the ith column of the decomposition of the covariance of the residuals, $\Sigma = A_0^{-1'} A_0^{-1}$. As noted earlier, typically this is a Cholesky decomposition of the error covariance matrix (a generalization of a square root for a matrix). This matrix A_0^{-1} provides an accounting of the contemporaneous correlations among the innovations. It is a lower triangular matrix, so the orthogonalized residuals then account for a particular pattern of linear combinations of the residuals.

We can use this orthogonalization to decompose the variance of the forecast errors with respect to time and each variable. We do this by replacing the residual covariance Σ in Equation 2.27 with the orthogonalized residual covariance of Equation 2.28. The orthogonalized forecast error variances in Equation 2.27 are then written as

$$V(y_{t+s} - \hat{y}_{t+s}) = \sum_{i=1}^{m} a_i a_i' + C_1 a_i a_i' C_1' \tag{2.29}$$
$$+ C_2 a_i a_i' C_2' + \cdots + C_{s-1} a_i a_i' C_{s-1}',$$

which is the moving average representation rescaled into the orthogonalized residuals. This variance matrix can be computed for each of s forecast horizon periods. The ith row of this matrix is the variance of that equation and the jth column is the covariance that is explained for the ith row variable. Thus, the elements of these s forecast variance matrices provide information about how much of the variation in each of the variables are explained by their own innovation and the variation in the innovations of the other variables in the system.

This decomposition of the forecast error variance is typically presented in a tabular or graph form by finding the percentage of a variable's forecast error variance that can be attributed to itself and the other variables in the system. In this table, the column variables account for a certain percentage of forecast error in the row variable at period s, where s is the number of steps after the innovation. A perfectly and strictly exogenous variable will have 100 for every s associated with its own column. Intuitively, this table provides information about how forecast errors would feed throughout a system. If one variable is important in accounting for the dynamics of another variable, then a forecast error in the first variable will lead to a forecast error in the second variable.

The forecast error decomposition for a VAR system is useful because it provides a way to assess the impacts of changes in the variables on each other. This can be a useful way of assessing Granger causal relationships and understanding how the series are related to each other. The two earlier interpretations of innovation accounting—analyzing forecast performance and ANOVA for a forecast horizon—are dynamic interpretations of the variance of the multiple time series in a VAR system. However, because the decomposition is computed in percentage terms, it does not produce an interpretation that is relative to the scales of the original variable. In addition, it does not provide a measure of the uncertainty of the variances associated with the impacts of shocks to the VAR system. IRFs or MARs can do this. Thus, one will often see innovation accounting and impulse response analysis in the application of VAR models.

2.6   Other Specification Issues

A number of issues arise in the specification and estimation of VAR models. The first of these involves the measurement and specification of the *variables* used in the model. Decisions about taking first differences of nonstationary series, data transformations (e.g., logarithms), and whitening are often considered. In addition, there are several other factors that can threaten VAR inferences. We address these issues in the final section of this chapter.

2.6.1 Should Differencing Be Used for Trending Data?

In univariate ARIMA modeling, transformations of the data are critical for specification analysis. Employing Box-Jenkins' technology is focused on determining whether or not a model is stationary and has white noise residuals. In VAR modeling, the presence of unit roots is also important. The requirements for estimation and the distributions of the parameters require covariance stationarity (Hamilton, 1994). Covariance stationarity is determined by having the roots of the *multivariate* lag polynomial for the VAR model outside the unit circle. This is hard to test for and evaluate (although methods are common in good time series software such as RATS). Instead, analysts often check the roots and look at the MAR to see if the shocks to the model die off or explode. Models with unit root variables can be directly estimated using vector error correction models (VECMs) and other methods. One should be aware that the VAR impulse responses will not be valid in cases where unit roots and trends are not correctly modeled (for a discussion, see Hamilton, 1994, chaps. 18 and 19).

What this suggests, then, is that data transformations that difference the data to remove trends should be avoided. In fact, VAR models can allow for complex error correction mechanisms and produce a stationary system with nonstationary individual series. In these cases, inferences about the dynamics in the data are possible asymptotically, with caveats (Freeman, Williams, Houser, & Kellstedt, 1998; Sims et al., 1990).

In addition, differencing the data in the VAR removes the long-term components of the series. Because these are often of primary interest in a dynamic analysis, differencing the data would eliminate a large portion of the trends and paths of the variables that we seek to explain.

2.6.2 Data Transformations and Whitening

What other transformations are typically employed in VAR models? A common transformation is to use the logarithms of trending or variance nonstationary variables. Employing logarithms of the data implies that the underlying VAR model is expressing a period-over-period growth rate for the data. This interpretation arises from the following approximation of growth rates (common in economics):

$$\frac{y_t - y_{t-1}}{y_{t-1}} \approx \ln(y_t) - \ln(y_{t-1}). \tag{2.30}$$

The left-hand side of this equation is the common way to measure period-over-period growth in a series. The right-hand side is its approximation. But this approximation is what is estimated in a VAR, because the endogenous variables at time t are on the left-hand side of the equation and the lagged values are on the right-hand side.

Using the natural logarithm transformation allows us to interpret the shocks in a MAR as percentage changes in the variables of interest. This is particularly useful for a model that might include a percentage variable (say, an interest rate or population growth) and a variable in levels (such as gross domestic product or some dollar-denominated figure). Working in logarithms of the variables allows us to interpret the percentage changes in both sets of variables in the MAR. An example of this type of data transformation and its interpretation is included in the subsequent examples.

Another possible transformation of the data that is often seen in some time series analyses is to prewhiten or filter the data. In this type of transformation, the data are put through a filter (either for deseasonalizing or smoothing the data). This should be avoided with data in VAR models. The reason is that this whitening or filtering alters the dynamics of the data and therefore the estimated relationships among the series in the system. In some cases, it is possible that such whitening can alter the dynamics and thus the assessment of Granger causality in the data (Sims, 1972).

2.7 Unit Roots and Error Correction in VARs

One of the alternative modeling strategies, suggested by the adherents of the LSE school, as discussed in Chapter 1, is error correction models (ECMs). An ECM specifies how two or more series of variables are related to one another via the short-run dynamics of the series and their long-run trends or equilibrium. Such models can be used for both stationary and nonstationary data, though most typically they are employed for modeling data with common stochastic trends. A unit root or stochastic trend variable is described as integrated to order d or $I(d)$ if it must be differenced d times to be made stationary. ECMs for multiple time series provide an explicit representation of the trend(s) in such series as well as the dynamics around these trends. The reason for these special models for unit root data is that there is a risk of spurious regressions—not correctly accounting for the trends in the data—that can lead to incorrect inferences (Granger & Newbold, 1974).

2.7.1 Error Correction Representation of Unit Root Data

This special application of ECM models to trending data is the result of the Granger representation theorem, which states that if two unit root variables share a common trend, then there exists a stationary linear regression model for the two series. This theorem is important for three reasons. First, it means that one can construct a model to evaluate whether two or more series share a common trend and dynamics without the risk of estimating a possibly spurious regression. Second, the theorem implies that there exists a Granger causality relationship among the two integrated or trending variables. This means that the innovations in one variable "drive" or cause the path of the second variable. A final implication of the theorem is that it allows for a description of the short- and long-run equilibrium dynamics of the integrated variables. This is accomplished by a regression model that accounts for the long-run (common trend) and short-run dynamics of the series (the error correction mechanism) (Engle & Granger, 1987).[12]

To see how ECMs allow us to see the short- and long-run dynamics, we first present a model for two unit root variables. If one were to model these variables with a VAR, the residuals could be nonstationary and inferences about the model parameters could have nonstandard distributions (not t, F, χ^2) (Toda & Phillips, 1993; Toda & Yamamoto, 1995). This is because there would be an unmodeled trend in the residuals. Thus, we need to consider how to recover information about both the trends and the short-term dynamics. To demonstrate how the ECM for $I(1)$ variables works, consider the following single-equation ECM for $I(1)$ variables Y_t and Z_t:

$$Y_t = \beta_1 Z_t + \beta_2 Y_{t-1} + \beta_3 Z_{t-1} + \epsilon_t, \tag{2.31}$$

where

$$\frac{\beta_1 + \beta_3}{1 - \beta_2}$$

is the total multiplier for the effect of changes in Z_t. This model will be cointegrated, meaning that there will be a linear combination of Z_t and Y_t that is stationary according to Granger's representation theorem (Engle & Granger, 1987). The total multiplier coefficients cannot be correctly estimated by OLS because of the nonstationarity of the residuals in Equation 2.31.

An error correction representation or version of the model can be derived from Equation 2.31 by subtracting out the first differences of the two variables from the model:

$$
\begin{aligned}
Y_t &= Y_{t-1} + \beta_1^* Z_t - \beta_1^* Z_{t-1} + \beta_2^* Y_{t-1} - \beta_2^* Y_{t-2} \\
&\quad + \beta_3^* Z_{t-1} - \beta_3^* Z_{t-2} + u_t, \\
(1-L)Y_t &= \beta_1^*(1-L)Z_t + \beta_2^*(1-L)Y_{t-1} \\
&\quad + \beta_3^*(1-L)Z_{t-1} + u_t,
\end{aligned}
\tag{2.32}
$$

$$\Delta Y_t = \beta_1^* \Delta Z_t + \beta_2^* \Delta Y_{t-1} + \beta_3^* \Delta Z_{t-1} + u_t. \tag{2.33}$$

The problem with the representation of the model in Equation 2.32 is that although one is working with the first differences or stationary data, we cannot recover the estimate of the total impact multiplier for Equation 2.31 (the unstarred coefficients). So as an alternative to Equations 2.31 and 2.32, Equation 2.33 can be made stationary by constructing an ECM as follows:

$$(1 - L)Y_t = \beta_1 Z_t + (\beta_2 - 1)Y_{t-1} + \beta_3 Z_{t-1} + v_t,$$
$$(1 - L)Y_t = \beta_1(1 - L)Z_t + (\beta_2 - 1)Y_{t-1} + (\beta_1 + \beta_3)Z_{t-1} + d_t,$$
$$\Delta Y_t = \beta_1 \Delta Z_t + (\beta_2 - 1)\left[Y_{t-1} + \left(\frac{\beta_1 + \beta_3}{\beta_2 - 1}\right)Z_{t-1}\right] + g_t, \tag{2.34}$$
$$\Delta Y_t = \beta_1 \Delta Z_t + (\beta_2 - 1)u_{t-1} + g_t.$$

Equation 2.34 is the error correction representation of the model. Estimating this final equation allows us to recover the estimates of the long-run impact multiplier (summarized by β_1 and β_3) and the short-term impacts (β_2)—even in the presence of $I(1)$ variables. The results will also be non-spurious, because the residual g_t will be stationary.

There are two methods of estimation of the ECM. One is a "two-step" method that estimates the long-run coefficients first (β_1 and β_3) and then estimates the error correction mechanism (β_2). This method will produce consistent, but possibly inefficient, estimates. The alternative "one-step" method employs only a single regression. In this method, one includes Y_{t-1} and Z_{t-1} in Equation 2.31. Estimating this new regression equation produces consistent and efficient estimates. Once estimated, the long-run component can be solved analytically. There is some evidence that the latter method works better in small samples.

2.7.2 Error Correction as a VAR Model

The single-equation ECM discussed in Section 2.7.1 is a simplification of a more general bivariate VAR model. If we explicitly specify a second equation for Z_t, we can analyze the possibility of cointegration within a VAR model. Let $y_t = (Y_t, Z_t)$ be a 1×2 vector of Y_t and Z_t. The VAR representation of this model is

$$y_t = \sum_{\ell=1}^{p} y_{t-\ell} A_\ell + u_t.$$

The same transformation that was used for the single-equation ECM can then be generalized to the VAR model. We first subtract y_{t-1} from the whole system and collect the terms in the model

$$y_t - y_{t-1} = -y_{t-1} + \sum_{\ell=1}^{p} y_{t-\ell} A_\ell + u_t,$$

(2.35)

$$\Delta y_t = y_{t-1}\Pi + \Delta y_{t-1}\Gamma_1 + \cdots + \Delta y_{t-p+1}\Gamma_{p-1} + u_t,$$

where

$$\Pi = -(I_m - A_1 - \cdots - A_p),$$
$$\Gamma_i = -(A_{i+1} + \cdots + A_p), \quad i = 1, \ldots, p-1,$$

and Δ represents the first difference of y_{t-k}, so that $\Delta y_{t-k} = y_{t-k} - y_{t-k-1}$. This model, which is a VECM of order $p-1$ or VECM($p-1$), allows one to recover both the long-term cointegrating relationship (Π) and the short-term impacts (Γ_i).[13] The reason this representation of the data is superior to cointegrated data is that it allows us to easily recover the cointegrating relationships in Π and the short-run impacts in Γ_i without any further transformations. If there are cointegration relationships in the model, then they are estimated in the long-run term Π. These cointegration relationships will be linear combinations of the variables, meaning that Π will not be full rank unless the number of cointegration relationships is selected first.

Estimation of this model requires the use of a set of reduced rank regression models for Equation 2.35 or a canonical correlation analysis (Johansen, 1995). This method will produce an estimate of the reduced rank matrix $\Pi = \alpha\beta$ that defines the cointegrating relationships. One can then substitute this estimate into Equation 2.35 and estimate the short-run effects Γ_i.

One question that must be answered in specifying this model is "how many cointegrating vectors or relationships are there?" For the above model with two possible unit root variables, there could be two independent random walks or a shared trend model with one random walk. This inference requires testing for the number of cointegrating vectors or the rank of the cointegration vector Π using modified likelihood ratio tests with nonstandard distributions. These tests are well described in standard texts and recent treatments such as Lutkepohl (2004) and Saikkonen and Lutkepohl (1999, 2000a, 2000b).

Each VECM($p-1$) model in first differences can be written as a VAR(p) in levels by transforming the VECM coefficients Γ and Π into the reduced form VAR coefficients A_i. The VECM coefficients in Equation 2.35 are related to the VAR coefficients via

$$A_1 = \Gamma_1 + \Pi + I_m,$$
$$A_i = \Gamma_i - \Gamma_{i-1}, \quad i = 2, \ldots, p-1,$$
$$A_p = -\Gamma_{p-1}.$$

As can be seen from this VAR representation of the VECM coefficients, the VAR "includes" all of the relationships in the VECM, but not in the same explicit notation. The autoregressive coefficients of the differenced VAR process can then represent the long-run relationships in the VAR just as effectively as the reduced rank estimated coefficients of the VECM. What one cannot do with the VAR coefficients is represent separately the short- and long-run effects modeled explicitly in the VECM. Because the VECM representation imposes a set of constraints on the relationship among the VAR coefficients, the VECM is typically seen as a VAR with long-run restrictions on the equilibrium relationships of the series.

2.7.3 VAR Versus VECM (ECM)

Why, then, would one select the VAR or VECM representation of the cointegrated model? The choice depends on whether or not one needs to estimate the cointegrating relationships or just the dynamic structure of the system of equations. If the central quantities of interest and analysis are the cointegration relationships, then one should use the VECM model. If, however, one wishes to assess only the causal relationships and short-term dynamics in the model, then the VAR approach can be used effectively. Using the VAR models for testing exogeneity restrictions when there is possible cointegration requires some care, because the presence of cointegration can lead to incorrect hypothesis tests (Sims et al., 1990). Modified tests for bivariate Granger causality in the presence of cointegration have been developed by Lutkepohl and Reimers (1992a) and Dolado and Lutkepohl (1996). Impulse response analysis of these models is addressed in Lutkepohl and Reimers (1992b).

VECM and ECM models are particularly appropriate for situations where there are multiple variables with possibly related trends. This is common in economics, where the underlying expansion of economic productivity drives many variables at the same time. There are several arguments that might lead one to reconsider the application of these models, broadly, in the social sciences as argued by Williams (1993). First, many variables that may appear to have stochastic and deterministic trends in a sample will not have these features in the long run. Second, the classical inference techniques as the basis of unit root econometric methods like VECM may be subject to criticism. Third, the error correction approach may have limited application in some fields.

First, tests for unit roots and error correction representations are conditional on the sample of data used to test for these properties. At first glance, unit root tests may appear to fail to reject the null of a unit root. Yet as argued forcefully by Williams (1993), a variable may have only a short-term trend (like presidential approval) and is not likely to be a unit root or explosive variable, because shocks to such a variable die off (slowly) over time. Thus, although in-sample data may often appear to be nonstationary, in practice, if theory or prior information suggests that these data are likely to be stationary, they should be modeled as such.

Second, testing for unit roots is a "knife-edge" exercise. The null hypothesis in standard augmented Dickey-Fuller (ADF) and Kwiatkowski-Phillips-Schmidt-Shin (KPSS) tests for a unit root is that the series has a unit root (Kwiatkowski, Phillips, Schmidt, & Shin, 1992). As Bayesians have noted, these tests tend to place too much probability on the likelihood of an explosive or nonstationary model (Sims & Uhlig, 1991). In the end, then, we are too likely to say that a series is a unit root using classical tests. As argued by Williams (1993, p. 231),

> Conceptually, a major difficulty with unit root tests is that it is quite possible that series exhibiting explosive or random walk patterns over finite sample periods will not indefinitely continue such a pattern. Classical inference is, as we all know, based on inferring something about a population from a sample of data. In time-series, the sample is not random, and the population contains the future as well as past. This alone should alert us to the danger of treating the battery of tests found in [research] as giving us precise probabilities over repeated samples. Lost in the roar of the countless numbers of unit root tests being performed in economics (and political science) is the fact that we treat the sample of time-series in a way that biases findings in favor of unit roots and trends. . . .
>
> Unit root econometrics seems most useful when testing for common trends in variables that will most probably continue trending throughout the future, such as consumption and income.

The third point made by Williams is that we should not "overgeneralize" the longevity or existence of common trends in data (Williams, 1993, p. 232). The real advantage of these models, then, is that they can be used to parsimoniously account for multiple time series that have shared short- and long-run dynamic processes—even if there is a question about the presence of a shared unit root process. VARs can be and often are profitably employed to describe error correction processes as well as more complex dynamics of large multiple time series systems. From the perspective of summarizing dynamics and describing relationships among multiple series, VAR models can and should be part of the approach. Such models should not exclude the possibility that a VAR can do as good a job explaining the process that generated the data as a VECM. One can easily investigate the long-run trends and cointegration properties of error correction relationships in a VAR because they are restrictions on the long-run behavior of a VAR model.

2.8 Criticisms of VAR

The main rationale for the VAR modeling approach is its sensitivity to identification assumptions. For those who find the identification assumptions of SEQ models to be untenable, VAR models are an attractive alternative approach. The main differences between VAR and SEQ modeling are the VAR focus on dynamics and the account of contemporaneous relationships via MARs.

Yet VAR models are subject to criticisms. Probably, the most vocal has been the charge of "overparameterization." Even small VAR models have a large number of regression parameters. One strain of criticism directed at VAR models (e.g., Pagan, 1987) argues that the large number of parameters in VAR models renders them so inefficient as to be nearly useless for inference. The criticism then goes on to say that F and χ^2 tests should be used to develop a more parsimonious model.

This argument is in large part a holdover from the earlier approach to specifying and testing regression and SEQ models. In these approaches, a general model is specified and then tested to determine if zero restrictions make sense. The goal is a more parsimonious model. The rationale for doing this is that the gain in degrees of freedom for the more parsimonious model will produce more efficient estimates and smaller confidence intervals for quantities of interest, including any forecasts, impulse responses, or policy analysis simulations.

VAR modelers recognize that the parsimony principle has significant power. In fact, this is to a large degree what makes ARIMA modeling so powerful, because the goal there is to find the most parsimonious representation of a variable's dynamics. But there are at least two reasons why VAR modelers reject this approach. First, how is one to make judgments about the significance of zero restrictions? Recall that these decisions are in a sense arbitrary, because they depend on the selection of a priori and possibly false restrictions to test and lead to underreporting of the inherent uncertainty of the model specification process. Although the resulting parsimonious model may appear more efficient, the reported efficiency of the final model is in fact false because it does not account for the model selection and specification processes.

VAR modelers also reject this "specify-estimate-test-respecify" logic as a means to model parsimony because it uses the data twice. Sims has argued that this is a serious problem because this becomes a source of overconfidence as the results seem more efficient than they really are (Sims, 1986b, 1988). The issue, though, with using the data this way (once for estimation and testing and a second time for respecification and estimation), is that it consumes the finite degrees of freedom twice. As such, the final estimates, forecasts, and inferences are too confident. This is a problem because it then means that forecasts and policy analysis are too confident and one is likely to be reporting confidence intervals and *P*-values that are in a sense "too good."

A second major criticism of VAR models is that they are "atheoretical." Cooley and LeRoy (1985) is probably the clearest example of this criticism. A main reason why the SEQ modelers object to the VAR approach is that it focuses on the reduced form rather than structural representation of the model. As discussed in Section 2.1 on the relationship between SEQ and VAR models and in Section 2.5.4 on Granger causality, noncausality in terms of a structural model does not mean the same thing as noncausality in the reduced form model. In large part, this criticism depends on the exogeneity restrictions in an SEQ model as a restriction on the identified parameters. VAR modelers, on the contrary, do not accept this view. Instead, they view the identification assumptions as distinct from those necessary to estimate the dynamics (under the assumptions of the Wold decomposition theorem). This criticism of the VAR approach then depends on the restriction or classification of variables as endogenous and exogenous. Although theoretically appealing, VAR modelers view such distinctions as generally dubious.

A third major criticism of VAR modeling and interpretation is based on Cooley and LeRoy's (1985) argument that VAR models are noncausal. VAR models and their interpretation via MARs and innovation accounting are based on "... identification of conditional correlations with causal orderings" (Cooley & LeRoy, 1985, p. 301). The rationale behind this criticism is that the conditional predictions about the effects of an endogenous variable, *X*, on another variable, *Y*, requires that *X* be weakly exogenous or predetermined with respect to *Y*. Otherwise, the feedback and related dynamics of the two variables cannot be determined (in terms of the structural model).

VAR modelers have several responses to this argument. First, it must be remembered that VAR models are based on the reduced form of the system of equations. As such, impulse response and innovation accounting analyses have the interpretation only as a causal or conditional prediction about the relationships among the variables *given the identification of the structural or reduced form error covariance*. That is, the interpretation of the VAR depends on the method and structure explicitly contained in the contemporaneous innovations.[14] If the ordering of the variables in a Cholesky decomposition of the error covariance matrix Σ is incorrect, then it is likely that the conditional predictions of the impulse response or innovation accounting will be incorrect.

Although exogeneity restrictions may not be a central part of VAR modeling, VAR modelers do not make the assumption that any exogeneity restrictions exist in the model *before* estimation. Working with the reduced form representation and then imposing causal orderings of the contemporaneous innovations in MARs and innovation accounting forces users of VAR models to be explicit about the assumptions they are making about the identification of the relationships among the variables. Furthermore, such decisions can be evaluated using different identification methods for the MAR and using Granger causality tests.

MARs and innovation accounting are based on tracing out the responses and changes in the variables in the system of equations for a *surprise shock or innovation*. In a reduced form VAR model with serially uncorrelated residuals, these shocks or innovations are exactly that—"surprises" or unpredictable random errors. This means that they are unpredictable with respect to the variables in the system and are therefore exogenous to the variables in the VAR system. Issues about the *contemporaneous* relationships among the variables remain, but the exogeneity of the innovations to each other can be established by considering the robustness of the MAR or innovation accounting to different identification assumptions as noted above.

VAR models are also a basic generalization of single-equation ARIMA, SEQ, and ECM models. In fact, one can view single-equation ARIMA, ECM, and SEQ models as special cases of a VAR (for further examples, see Reinsel, 1993). Thus, the basic argument for VAR models is that they allow one to have a general model with fewer chances of making incorrect restrictions that can bias dynamic inferences.

A final response of VAR modelers to these critiques has been the development of additional extensions to the basic model presented here. In part, modifications of the basic VAR models allow for more parsimonious models and structural identification as in SEQ models. But rather than test explicitly for the restrictions, these extensions to the basic VAR model presented here approach the restrictions using Bayesian methods that place "loose" or probabilistic restrictions on the model parameters. A second vein of extensions to VAR models has developed structural interpretations of the contemporaneous causal relations, known as *structural VAR* or SVAR models. More recent work has combined these two developments into Bayesian SVAR models (e.g., Leeper, Sims, & Zha, 1996).

3. EXAMPLES OF VAR ANALYSES

This chapter presents two complete examples of vector autoregression (VAR) or multiple time series analysis. In each example, we discuss how a VAR model can be specified, estimated, and interpreted. Our main goal is to provide a clear illustration of the models we discussed in Chapter 2. These examples are intended to serve as straightforward demonstrations of VAR models and their inference. We view these examples in the spirit of responding to the criticism of Pagan (1987), who argued that the methodology and decisions that go into VAR models make them hard to evaluate and interpret.

The first example presents an analysis of the dynamics of American political partisanship and the public's policy preferences for government action. This is a model of the relationship between Stimson's (1999) measure of the "public mood" and aggregate partisanship or macropartisanship. This is a simple example that allows us to illustrate some of the basic uses of VAR models. Also, we readily admit that this model is likely underspecified, but it does allow us to illustrate the basic ideas.

The second example is a more complicated example from Williams and Collins (1997) about the impacts of political and economic factors on effective corporate tax rates in the United States. This empirical example is built on a theoretical rational expectations model of the relationship between business interest groups' political influence, economic conditions, and tax rates on businesses. This example is much more developed, but this means that the analysis is more complex.[1]

3.1   Public Mood and Macropartisanship

How is the American public's aggregate support for government and its actions related to the aggregate partisan identification of the American public? This question, common in the literature on American public opinion, is one that involves two endogenous concepts. One expectation is that as the public desires more government activism or approves of the activism, they are more likely to identify with the party in power. Thus, changes in the public mood should be tied to aggregate partisan identification. These issues have been explored in detail by several authors, including Stimson (1999) and Erikson, MacKuen, and Stimson (2002).[2]

There are two variables used in this example. The first is the *public mood*, which measures the general public support for government action in the United States. This measure is created through an aggregation of a series of Gallup poll questions using a dynamic factor model (Stimson 1999). Theoretically, this measure is based on the percentage of Americans who support

60

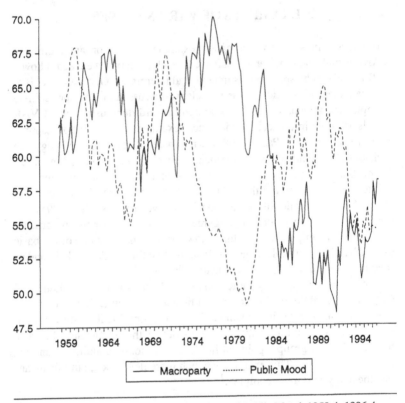

Figure 3.1 Quarterly Macropartisanship and Public Mood, 1958:4–1996:4

government policies across a series of polls over time. The measure aggregates multiple survey questions over time and can theoretically range from 0 to 100, with higher values indicating more support for government action and lower values indicating less support for government action. The second variable, *macropartisanship*, is the measure of the Democratic share of the two-party identification share based on Gallup poll surveys. In our example, we look at quarterly data from the fourth quarter of 1958 to the fourth quarter of 1996.

Figure 3.1 plots the two data series. Note that both series appear to have local trends—for short stretches of time the series appear to move in a specific direction. Therefore, in addition to testing for the lag length of the VAR, tests should also be conducted for unit root or trends. Three control variables will also be included for the partisan control of the White House, U.S. House of Representatives, and the U.S. Senate and these are coded

"1" if the branch of government is controlled by the Democratic Party and "0" if the branch is controlled by the Republican Party.

The specification and analysis of these series using a VAR model follow the steps outlined in Chapter 2:

1. Test for unit roots.
2. Test for lag length.
3. Specify and estimate the VAR model.
4. Test for Granger causality.
5. Decompose the forecast error variance for innovation accounting.
6. Conduct and interpret an impulse response analysis.

We present each of these steps in detail to show how we arrive at a model specification for the VAR. We then illustrate how to interpret the VAR estimates and describe the dynamic relationships among the public mood and macropartisanship variables.

3.1.1 Testing for Unit Roots

We first test for unit root behavior in the macropartisanship and public mood variables. Testing for unit roots is important because if one or more of the variables has a unit root, there is the possibility that there is an error correction mechanism and that tests for Granger causality may be suspect. Testing for unit roots is done using ADF tests (for details on these tests, see Hamilton, 1994). Table 3.1 shows the results for the ADF tests augmented for lags 0 to 8.

The critical value for the ADF tests for the null that there is a unit root for a model with a random walk and a drift (a constant) is approximately −2.88. We see that for the first column of Table 3.1, there is evidence that there is a random walk in the macropartisanship variable—the test statistics are less than the critical value under the null of a unit root (consistent with the findings of Box-Steffensmeier & Smith, 1996). However, for the case with a trend, there is good evidence to fail to reject the null—but in these cases, the estimated trends are not statistically significant. For the public mood variable, the deterministic trend is insignificant, and the ADF test indicates that the null of a unit root could possibly be rejected for models with more lags.

It would appear that there may be a unit root in the macropartisanship variable. Yet this variable is bounded above and below—it is the percentage of respondents in Gallup polls who say that they are Democrats relative to the total number of partisans (Democrats plus Republicans). This variable, then, cannot go outside the range of 0 to 100, and is therefore not likely to

TABLE 3.1
Augmented Dickey-Fuller Test Results

	Macropartisanship		Public Mood	
Lags	No Trend	Trend	No Trend	Trend
0	−2.04	−2.97	−1.75	−1.92
1	−1.65	−2.33	−1.87	−2.03
2	−1.55	−2.24	−1.93	−2.09
3	−1.60	−2.46	−2.08	−2.19
4	−1.79	−2.73	−2.49	−2.59
5	−1.54	−2.45	−2.67	−2.75
6	−1.48	−2.22	−2.57	−2.55
7	−1.30	−2.25	−2.31	−2.28
8	−1.08	−2.10	−2.82	−2.79

NOTE: Critical value at the 5% level for the no trend test is approximately −2.88. Critical value at the 5% level for the trend test is approximately −3.50.

be nonstationary over longer time horizons. This fact can be reconciled with the finding of a possible unit root because the *underlying* concept of aggregate partisanship may be nonstationary over time or over short periods of time. It may be a process with local trends and a drift. However, over the longer term, it will be stationary—once the aggregate percentage of Democrats reaches a high percentage, it must fall. Thus, theoretically this variable should be treated as stationary as suggested by Williams (1993).[3]

3.1.2 Specifying the Lag Length

Specification of the lag length for the VAR is typically done using a combination of fit statistics and formal test statistics for lag length. One should remember, though, that any finite order lag length is only an approximation to a (possibly) infinite-order lag length. Therefore, in applied work, one should take a holistic approach looking at lag tests and fit statistics.

Table 3.2 presents the Akaike information criterion (AIC) and Bayesian information criterion (BIC) values for VARs with 1 to 12 lags. These should be interpreted as fit statistics that describe the improvement in the log-likelihood, penalized for the additional lags. Smaller values of these fit statistics are better (because they are based on the negative of the log-likelihood). Based on this table, one could choose a lag length of $p = 1$. A case could be made for 2 lags, because the difference in the values of AIC and BIC are very close for 1 versus 2 lags (compared to 2 vs. 3). If one were worried about residual serial correlation, then adding additional lags only brings an efficiency penalty (not a bias penalty) in the estimates. In what

TABLE 3.2
AIC and BIC Values for
Macropartisanship and Public Mood VAR

Lags	AIC	BIC
1	238.24	255.58
2	240.86	269.77
3	246.69	287.15
4	253.75	305.78
5	251.26	314.85
6	252.95	328.10
7	259.53	346.24
8	259.11	357.38
9	255.47	365.30
10	256.07	377.47
11	260.29	393.24
12	263.00	407.52

follows, we will use only 1 lag. Notice that the changes in the likelihood are nearly the same for each lag added. As noted in Chapter 2, each additional lag incurs a penalty of $2m^2 = 8$ for the AIC and $\log(T)m^2 = \log(124)2^2 = 8.37$ for the BIC. Thus, changes in the AIC and BIC that are less than this value are not likely to have a sufficient improvement in model fit to justify the inclusion of another lag.[4]

An additional method for evaluating the lag length can be the use of hypothesis tests for lag length. Table 3.3 shows a similar conclusion. The χ^2 test presented here is for the addition of one more lag versus a null of one fewer lag. Because there are two variables, each new, there is weak evidence for not rejecting the 2-lag model in favor of the 1-lag model (P-value $= 0.27$), but there is more conclusive evidence for rejecting the 3-lag model in favor of the 2-lag model (P-value $= 0.72$). Based on these results, the subsequent VAR model will use only 1 lag. This decision presents a situation that one would want to check by running models with both 1 and 2 lags.[5]

3.1.3 Estimation of the VAR

In this analysis, we use a 1-lag VAR model. We also include in each equation three control variables and an intercept. The control variables are dummy variables for the party in control of the presidency, Senate, and House of Representatives. These are coded "0" in years that Republicans control the office or chamber and "1" in years that the Democrats control the office or chamber.

TABLE 3.3
Likelihood Ratio Tests for Lag Length

Unrestricted Lag Length	Restricted Lag Length	χ^2	P-value
12	11	4.29	0.37
11	10	3.13	0.54
10	9	6.23	0.18
9	8	9.98	0.04
8	7	7.35	0.12
7	6	1.26	0.87
6	5	5.70	0.22
5	4	9.62	0.05
4	3	0.88	0.93
3	2	2.06	0.72
2	1	5.18	0.27

NOTE: Test statistics are χ^2 with 4 degrees of freedom.

TABLE 3.4
VAR Estimates for the 1-Lag Model of Public
Mood and Macropartisanship

Variable	Dependent Variables	
	Public Mood	Macropartisanship
Public Mood$_{t-1}$	0.939 (0.029)	0.013 (0.042)
Macropartisanship$_{t-1}$	−0.019 (0.022)	0.954 (0.031)
President$_t$	−0.424 (0.282)	−0.041 (0.409)
House$_t$	0.077 (0.614)	−1.069 (0.889)
Senate$_t$	−0.187 (0.348)	0.410 (0.504)
Constant	4.900 (2.233)	2.743 (3.232)
R^2	0.919	0.897
Standard error	1.290	1.867
Durbin-Watson	1.952	2.381

NOTE: Standard errors in parentheses.

The model has six coefficients per equation with control variables for the party of the president, partisan control of the House of Representatives, and the Senate (1 lag × 2 variables plus a constant and three control variables). Table 3.4 presents the VAR coefficients and their standard errors for the two equations.

Typically, VAR analyses do not present these coefficients. The reason becomes apparent on review of Table 3.4. The autoregressive coefficients

TABLE 3.5

Granger Causality Tests for Public Mood
and Macropartisanship Based on the VAR(1) Model

Hypothesized Exogenous Variable	Block Coefficients Restricted	F Statistic	P-Value
Public mood	Macropartisanship	0.74	0.39
Macropartisanship	Public mood	0.09	0.76

in the first four rows of the table describe the dynamics of the system. These coefficients mean nothing individually, because it is the behavior of the system and all its coefficients that describe the dynamics of the variables. Any inferences about these parameters involve the four coefficients for the first and second lags. Because of the system dynamics, it is the matrices of VAR coefficients that need to be used to evaluate stationarity.

We can and do interpret here the control variables—the party in control of the presidency, House, and Senate—for their signs. These estimated coefficients describe the direction of the party control of the presidency or congressional chamber on the equilibrium level of each endogenous variable. Only the coefficients for the control of the presidency in the public mood equation ($t = 1.5$) and the control of the House in the macropartisanship equation ($t = -1.2$) approach statistical significance. We see that Democratic control of the presidency may depress the average public mood and Democratic control of the House also depresses aggregate Democratic partisan identification. These estimated impacts on the level of macropartisanship and public mood are complicated by the dynamics of the system. Although the signs can be interpreted, the magnitude of changes in these exogenous variables depends on the system dynamics. We are more interested in these dynamics and view these effects merely as control for historical factors. It is for these reasons that alternative methods such as decomposition of the forecast errors and impulse responses are used to interpret VAR models and engage in inference.

3.1.4 Granger Causality Testing

The VAR estimates in Table 3.4 and the 2×2 covariance matrix of the residuals can be used to assess whether there is Granger causality present in the system. As discussed in Chapter 2, Granger causality can be assessed using F tests in a VAR(p) model. Here, we use 1 lag, consistent with the earlier results.

Table 3.5 presents the results of the F tests from the VAR for Granger causality. Here, the test is whether the coefficients for the lagged public mood variable are zero in the macropartisanship equation and whether the coefficients for the lagged macropartisanship are zero in the public mood equation.

For both the exogeneity tests, we fail to reject the null hypothesis that the coefficients are zero. Thus, there appears to be evidence that public mood does not depend on the past values of the macropartisanship and that macropartisanship does not depend on the past values of public mood. This would appear to be a major problem—because if this conclusion holds, then the dynamics of the public mood and macropartisanship are mainly related contemporaneously. However, Granger causality tests are biased against the null hypothesis when there is a possible unit root (Hamilton, 1994, p. 554).

One way to see this is to recall the VAR estimates in Table 3.4. Here, we see that the VAR autoregressive coefficient matrix approaches an identity matrix. This is the matrix analog of finding a unit root. One way to see if this possible unit root behavior is the source of the bias in the Granger causality results would be to compare the results of Granger causality tests with different lag lengths. This is because the Granger causality test is sensitive to lag length, as are the unit root results. Doing so does not change the results: For any lag length from 1 to 8, the same conclusions as those from Table 3.5 hold.

3.1.5 Decomposition of the Forecast Error Variance

Forecast error decomposition is based on determining how much the fitted model differs from the actual values of the vector of endogenous variables. This is done by using the vector moving average interpretation of the VAR to compute the forecast deviations over different time horizons. The variance of these forecast errors is then decomposed and the percentage of the forecast variance due to each endogenous variable is determined. If the variables are exogenous of each other, we expect to see that the innovations in one variable do not explain the variation in the other variable. If the variables are contemporaneously correlated, we expect to see that the variation in one variable can begin to explain the other with a lag as the contemporaneous innovations work through the lags in the system of equations.

The forecast error decomposition for the VAR model of public mood and macropartisanship is presented in Table 3.6. The first two columns show the proportion of the forecast variance for the public mood variable from innovations or shocks to public mood and macropartisanship. From the discussion in Chapter 2, this decomposition depends on the correlation in the errors in the two series. This correlation is based on the covariance matrix of the residuals from the VAR. Because the public mood variable is first in the ordering of the variables, the decomposition assumes that the initial period has all the variance in the forecasts attributed to public mood and none to the macropartisanship measure. As the forecast horizon increases,

TABLE 3.6

Decomposition of the Forecast Error Variance for
the VAR(2) Model of Public Mood and Macropartisanship

	Forecast Error % for Public Mood Innovations In		Forecast Error % for Macropartisanship Innovations In	
k	Public Mood	Macropartisanship	Public Mood	Macropartisanship
1	100.000	0.000	2.755	97.245
2	99.963	0.037	2.904	97.096
3	99.878	0.122	3.052	96.948
4	99.750	0.250	3.197	96.803
5	99.582	0.418	3.340	96.660
6	99.376	0.624	3.480	96.520
7	99.136	0.864	3.617	96.383
8	98.866	1.134	3.751	96.249
9	98.570	1.430	3.881	96.119
10	98.251	1.749	4.008	95.992
11	97.913	2.087	4.130	95.870
12	97.559	2.441	4.249	95.751
13	97.192	2.808	4.364	95.636
14	96.817	3.183	4.475	95.525
15	96.435	3.565	4.581	95.419
16	96.049	3.951	4.684	95.316

there is more variation attributed to the other innovations based on the correlation of the innovations and the dynamics of the system. In this case,
after 10 quarters or 2.5 years, about 1.75% of the forecast variation in public
mood can be attributed to innovations in macropartisanship. This pattern is
roughly stable after this point, so that by the 16th quarter, 3.95% of the forecast error in public mood is due to innovations in macropartisanship.

Columns two and three in Table 3.6 are the forecast error decompositions
for macropartisanship. After 16 quarters, approximately 5% of the forecast
error in macropartisanship is attributable to innovations in public mood.[6]

The conclusion from this forecast decomposition analysis is that the
unexpected changes in macropartisanship have a small effect on the innovations in public mood—no more than 4% of the variance. Conversely, innovations in public mood also have small relative impacts in predicting the
forecast variation in macropartisanship. Substantively, we see that both the
public mood policy measure and macropartisanship have weak and slow
responses to each other. This is as would be expected: Preferences lead to
policy choice. This conclusion is robust to reordering the variables in the
Cholesky decomposition of the error covariance used to compute the
decomposition of the forecast error variance percentages in Table 3.6.

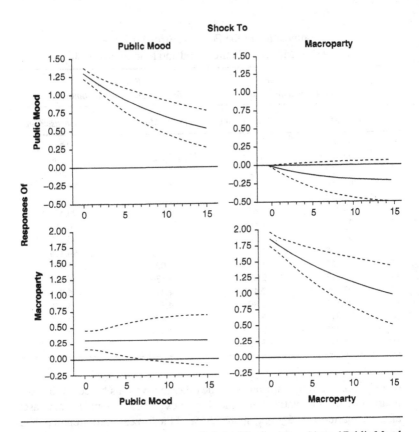

Figure 3.2 Impulse Response Analysis for Macropartisanship and Public Mood

NOTE: Public mood is first in this ordering. Error bands are approximate 68% confidence intervals or 1 standard deviation.

3.1.6 Impulse Response Analysis

Impulse response analysis allows an analysis of the dynamics of a VAR model in its vector moving average (VMA) representation. Substantively, this allows us to trace out the dynamic impacts of changes in each of the endogenous variables over time. One of the issues in the specification of the VMA process and the subsequent impulse response analysis is the ordering of the contemporaneous correlations. Here, both orderings of the contemporaneous correlations—based on a Cholesky decomposition of the estimated residual covariance matrix—are presented.[7]

Figure 3.2 presents the impulse responses with public mood as the first variable in the ordering of the decomposition of the contemporaneous covariance of the residuals. Note that this means that the shock to the macropartisanship is initially zero. So in the first row, the responses of public mood to the two shocks are presented. The approximately 1.25 unit shock (1 standard deviation of the public mood equation residuals) to public mood decays by about one half after 16 quarters. The upper right graph in the figure shows the response of public mood for a shock in macropartisanship. Here, the responses are presented with approximately 1 standard deviation or 68% confidence intervals. As noted in Chapter 2, these provide a better summary of the central tendency of the responses, as suggested by Sims and Zha (1999). The response of public mood to innovations in macropartisanship is generally weak and the 68% confidence interval includes zero for all 16 quarters.

The second row of graphs in Figure 3.2 shows the responses of macropartisanship to shocks in the two variables for the same ordering. Here, 1.25 unit shock to public mood leads to a sustained 0.25 point increase in macropartisanship for six quarters or 1.5 years. This effect becomes insignificant later, as the confidence region encompasses zero. The final, bottom right graph in Figure 3.2 shows the results of a shock to macropartisanship to itself. Because of the strong persistence in this variable and its unit root-like behavior, the impacts of this shock persist over many quarters and decay slowly.

These results illustrate two facts about the dynamic responses of macropartisanship and public mood. First, the response of public mood to changes in macropartisanship is rather weak. Conversely, macropartisanship has a sustained weak response to public mood. Second, these results are driven in large part by the modest contemporaneous correlation in the shocks to the two equations in the system. The contemporaneous residuals used to compute the residual covariance matrix are correlated at 0.17.

The choice of the ordering of the variables in the impulse responses is, however, not derived from political theory. As a result, the alternative ordering where macropartisanship is placed first in the decomposition should be analyzed. Figure 3.3 examines this alternative ordering of the impulse responses.

In Figure 3.3, the responses of macropartisanship and public mood to their own shocks—the responses on the main left-right diagonal of the matrix of graphs—are identical to those in Figure 3.3. What changes, however, are the responses to shocks in the variables on the right-left off-diagonal. In this second ordering of the decomposition, shocks to macropartisanship lead to very small and quickly statistically insignificant responses in public mood (the

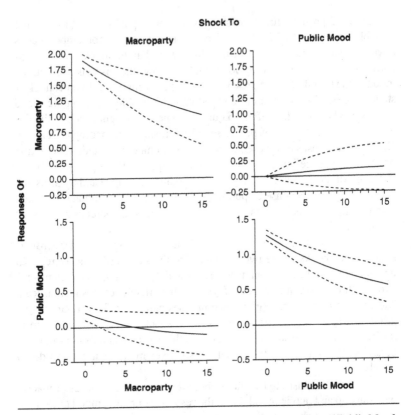

Figure 3.3   Impulse Response Analysis for Macropartisanship and Public Mood

NOTE: Macropartisanship is first in this ordering. Error bands are approximate 68% confidence intervals or 1 standard deviation.

bottom left graph in Figure 3.3). Shocks to public mood lead to a modest, although statistically insignificant, increase in public mood (based on the 68% error bands).

In conclusion, the ordering of the variables in the impulse responses appears to have little impact on the conclusions. If one believes that public mood shocks precede those to macropartisanship, then there is approximately a 0.25 point change in macropartisanship for 1 standard deviation shocks to public mood. In contrast, if one believes that shocks to macropartisanship precede those to public mood, then the response is smaller. In both cases, there is evidence that the variables are contemporaneously related and that shocks in one variable lead to a weak and slow dynamic response in the other.

3.2 Effective Corporate Tax Rates

Our second example is a more complex VAR analysis. It includes richer dynamics, unit root variables, a more theoretical specification, and inferences about the causal relationships among the variables. In this example, we replicate part of the analysis of Williams and Collins (1997). They use a VAR model to analyze the determinants of corporate tax rates in the United States. Some would argue that tax rates respond to political pressures. Republican control of the Congress and the White House should then lead to efforts by corporate political action committees (PACs) to put pressure on members of Congress to lower tax rates on business. Alternatively, Williams and Collins argue that a political economy model of tax policy predicts that optimal tax policy should be exogenous of political variables but respond to economic variables.

Williams and Collins (1997) present an optimal taxation model that predicts that political variables such as PAC contributions and business PAC size relative to other PACs should be exogenous to changes in effective corporate tax rates (ECTR). Their theory predicts that business's need to plan for investment decisions and the state of the economy are at the root of tax policy. Thus, tax policy manipulation should be determined by economic fundamentals and not by the political pressures of business. Their model generates three propositions about the relationship between corporate tax rates, the number of corporate PACs, and the state of the economy.

Proposition 1 The ECTR will be exogenous to the political strength of business interests in the aggregate. Rather, an increase in the ECTR will lead to an increase in business political organizational activity.

Proposition 2 Investment is sensitive to the ECTR: A shock in effective tax rates will reduce levels of investment.

Proposition 3 The ECTR will be exogenous to economic conditions, including aggregate real investment and real income.

They test this theory using a four-variable VAR model. The first proposition is a prediction about the direction of Granger causality. ECTR is hypothesized to be exogenous of the past political strength of business, but changes in ECTR are hypothesized to cause changes in future business strength. Proposition 2 states that the response of a shock to ECTR is a decrease in investment. This means that the impulse response and decomposition of forecast error variance for this effect should demonstrate this behavior. Finally, the past values of the economic variables should not

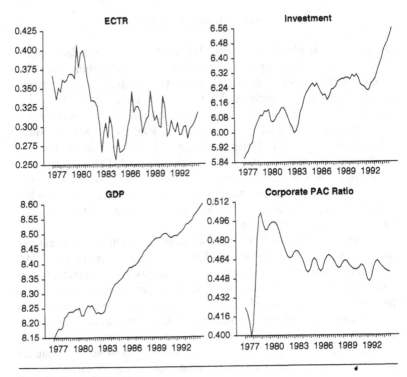

Figure 3.4   ECTR Data

predict contemporaneous values of ECTR—which can be demonstrated via Granger causality tests.

3.2.1   Data

Williams and Collins (1997) use data from the first quarter of 1977 until the last quarter of 1994 for their key results. This is because PAC data are available only after the campaign finance reforms and the beginning of campaign finance reporting in 1977. The four variables in their analysis are effective corporate tax rates (*ECTR*), the natural logarithm of real gross domestic product (*GDP*), the natural log of real investment spending (*Real Investment*), and the ratio of corporate PACs to all PACs (*Corp. PAC Ratio*). Figure 3.4 shows the graphs of the data.

For these data, one needs to consider the possible trends in specifying a VAR model. The data in Figure 3.4 likely contain unit roots, particularly the real investment and real GDP measures. The next two sections examine

TABLE 3.7
Unit Root Tests for ECTR Example Variables

Lags	ECTR	GDP	Investment	PAC Ratio
1	−2.92145	−0.56335	−0.42694	−5.11362
2	−2.78095	−0.71887	−0.40087	−3.60019
3	−2.55360	−0.95425	−0.48052	−5.03627
4	−2.06522	−1.09437	−0.61641	−3.99344
5	−1.92103	−1.20439	−0.59058	−1.72695
6	−1.93909	−1.06926	−0.46436	−2.18983
7	−1.86593	−1.03568	−0.61092	−3.22650
8	−1.65129	−0.83706	−0.54889	−3.13094

NOTE: Results are ADF test results.

the presence of unit roots and the lag length specification for the proposed four-variable VAR model.

3.2.2 Testing for Unit Roots

Economic variables such as GDP typically have a unit root. One concern in the present data is whether or not there are unit roots that would imply a cointegrating relationship among the variables. Table 3.7 presents the results of the ADF test for each series in the model.

Using the −2.88 critical value for these tests, we see that we can reject the unit root for the ECTR and PAC Ratio variables (as one should probably expect). The GDP and Investment variables, however, do contain unit roots. The implication of this is that interpretation of test statistics should be conservative and that there may be a cointegrating relationship in the data.[8]

Even with this evidence and the possibility of a cointegration relationship, we do not choose to present the error correction or vector error correction model. This is for two reasons: First, we expect that if there is cointegration among the GDP and investment variables, the resulting system will be stationary and thus render any test statistics asymptotically valid. Second, the cointegration relationship is not the central focus of the model's propositions. The short-term dynamics and exogeneity analyses are the main focus. Inferences about these dynamics and exogeneity will be robust in this situation.

3.2.3 Specifying the Lag Length

Given that the data are quarterly, it is common to use a lag length of 6 to account for seasonality in the data. In much of the following analysis, a lag

TABLE 3.8
AIC and BIC Lag Length Diagnostics
for the ECTR VAR Model

Lag	AIC	BIC
1	−2220.07	−2178.19
2	−2298.53	−2223.13
3	−2315.44	−2206.53
4	−2322.42	−2180.00
5	−2315.78	−2139.85
6	−2349.06	−2139.63
7	−2345.59	−2102.65
8	−2403.67	−2127.22
9	−2442.30	−2132.34
10	−2466.42	−2122.95
11	−2500.43	−2123.45
12	−2548.22	−2137.73

length of 6 will be used. To see if this is correct, Table 3.8 presents the AIC and BIC statistics for VAR models with lags between 1 and 12.

As seen in Table 3.8, starting at lag 1, the AIC drops until 4 lags, then increases at 5 lags, and drops again at 6 lags. Because these first "low" values are at 4 and 6 lags, this is evidence for using these lag lengths. Similarly, the lowest BIC values are at 2, 3, 4, 1, and 6 lags, respectively. Because the differences in the 4 and 6 lag values are not far from the lowest information criteria values, one should probably consider the longer-lagged models to ensure that there is no residual serial correlation.

Alternatively, one might look at the results of likelihood ratio tests for the lag length. Table 3.9 presents the results of likelihood ratio tests for successive lag lengths. It should be noted, however, that these results are not likely to be robust because the presence of unit roots in some of the variables will invalidate classical hypothesis testing for lag length. Working sequentially from 1 to 2, then 2 to 3 lags, and so on in Table 3.9, one would first reject a lag restriction at 4 lags, then again at 6. This is consistent with the AIC and BIC results, so we will investigate models with both 4 and 6 lags. This is consistent with the fact that we are using quarterly data and the fact that a 6-lag model will capture the residual seasonal effects in the data.[9]

3.2.4 Granger Causality Testing

The hypotheses about the ECTR, business PAC strength, and the state of the economy are concerned with Granger causality relationships. Bivariate Granger causality is assessed for the key variables using a series of bivariate

TABLE 3.9
Likelihood Ratio Tests for Lag Length

Unrestricted Lag Length	Restricted Lag Length	χ^2	P-Value
12	11	14.63	0.55
11	10	16.50	0.42
10	9	17.77	0.34
9	8	27.07	0.04
8	7	40.53	< 0.01
7	6	14.74	0.54
6	5	38.08	< 0.01
5	4	16.48	0.42
4	3	27.94	0.03
3	2	38.31	< 0.01
2	1	93.89	< 0.01

NOTE: Test statistics are χ^2 with 16 degrees of freedom with small-sample correction.

VAR models.[10] Table 3.10 presents the results of the exogeneity tests for these hypotheses.

Table 3.10 shows the results of exogeneity tests for the ECTR and PAC measures. The table presents exogeneity tests for the levels and first differences of the ECTR and various PAC measures (denoted by Δ). This is because there is the possibility that there are unit roots in the ECTR measure. The first column shows the variable that is hypothesized to be exogenous, and the second column shows the variable whose coefficients are restricted to zero under the null of Granger noncausality. Insignificant F statistics and P-values in the final two columns are evidence that one cannot reject the null hypothesis of noncausality.

In the main, these F statistic values indicate that one cannot reject the null hypothesis of noncausality. The tests show that ECTR is exogenous or not Granger caused by the number of and ratio of corporate PACs. Using a 0.05 level of significance, ECTR is exogenous to the ratio of corporate PACs. The only cases where this result is questioned is where the total number of business PACs is used in levels (P-value = 0.18 for the 4-lag models). Because it may be the case that the test statistic will be biased against the null hypothesis of exogeneity due to the trend in the number of PACs, one should also look at models with different lag lengths. The results of the 6-lag models confirm the results of the 4-lag models.

The test results in Table 3.10 also support the idea that ECTR Granger causes the ratio of corporate PACs. We see that we can reject the null of noncausality for the effect of ECTR on the ratio of corporate PACs (P-value = 0.01 for the 4-lag model and P-value = 0.12 for the 6-lag model). Thus,

TABLE 3.10

Exogeneity Tests for ECTR and Corporate
Political Action Committees, 1977–1994

Hypothesized Exogenous Variable	Block of Coefficients Restricted	F Statistic	P-Value
Four-quarter lag structure results			
ECTR	Number of Corporate PACs	1.62	0.18
Number of Corporate PACs	ECTR	0.31	0.87
ECTR	Corporate PAC Ratio	0.40	0.81
Corporate PAC Ratio	ECTR	9.51	< 0.01
ΔECTR	ΔNumber of Corporate PACs	0.42	0.79
ΔNumber of Corporate PACs	ΔECTR	0.43	0.78
ΔECTR	ΔCorporate PAC Ratio	0.36	0.84
ΔCorporate PAC Ratio	ΔECTR	0.35	0.84
Six-quarter lag structure results			
ECTR	Number of Corporate PACs	0.97	0.45
Number of Corporate PACs	ECTR	0.50	0.80
ECTR	Corporate PAC Ratio	1.86	0.11
Corporate PAC Ratio	ECTR	1.79	0.12
ΔECTR	ΔNumber of Corporate PACs	0.83	0.55
ΔNumber of Corporate PACs	ΔECTR	0.21	0.97
ΔECTR	ΔCorporate PAC Ratio	0.83	0.55
ΔCorporate PAC Ratio	ΔECTR	1.44	0.21

NOTE: Results based on bivariate VAR models.

changes in ECTR lead to subsequent changes in the activism of business. This directly supports Proposition 1. ECTR does not Granger cause the number of total business PACs—consistent with the growth of PACs over time. The trend may invalidate the F statistic. But the measures using the ratio of corporate PACs are consistent with Proposition 1.

Table 3.11 presents the exogeneity tests for ECTR and logged real investment and logged real income. The presentation is the same as that of Table 3.10. The results of these tests for various lag lengths lead to the conclusion that one should fail to reject the null hypothesis that the VAR coefficients for real income and real investment are zero. Using levels and differences to account for the possible unit roots in the data, it appears that ECTR is exogenous to investment and income. This supports Proposition 3, because ECTR is exogenous to the state of the economy.[11]

The exogeneity tests in Table 3.11 should not be used to test Proposition 2—that ECTR changes cause investment. This is because in two-variable VARs, incorrect or spurious findings of causality are likely when testing causality versus exogeneity. Proposition 2 is a statement about causality and not exogeneity, and thus depends on the dynamics of the variables in

TABLE 3.11

Exogeneity Tests for ECTR, Real Investment,
and Real Income: 1953–1994, 1960–1994, 1977–1994

Hypothesized Exogenous Variable	Block of Coefficients Restricted	Four Lags		Six Lags		Eight Lags	
		F	P	F	P	F	P
1953–1994							
ECTR	Real investment	1.27	0.28	1.05	0.40	1.15	0.33
ECTR	Real income	1.77	0.14	1.44	0.20	1.82	0.08
ΔECTR	ΔReal investment	1.02	0.40	1.12	0.35	1.02	0.42
ΔECTR	ΔReal income	0.85	0.50	1.59	0.15	1.98	0.05
1960–1994							
ECTR	Real investment	0.72	0.58	0.82	0.56	0.91	0.51
ECTR	Real income	1.13	0.35	1.24	0.29	1.54	0.15
ΔECTR	ΔReal investment	0.80	0.53	0.96	0.45	0.87	0.54
ΔECTR	ΔReal income	0.76	0.55	1.49	0.19	1.65	0.12
1977–1994							
ECTR	Real investment	0.51	0.73	0.36	0.90	0.76	0.64
ECTR	Real income	0.10	0.99	0.46	0.84	1.06	0.41
ΔECTR	ΔReal investment	0.68	0.61	0.52	0.79	0.56	0.80
ΔECTR	ΔReal income	0.28	0.89	0.95	0.47	1.35	0.24

NOTE: Results based on bivariate VAR models.

the system. The subsequent sections analyze these dynamics and assess the predictions of Proposition 2 using other methods.

3.2.5 Impulse Response Analysis

Sections 3.2.3 and 3.2.4 have established the basic exogeneity relationships of the variables in the model and the presence of possible unit roots. To determine the dynamics of the system of equations, we employ the same four-variable VAR as in Section 3.2.4. In this section, we use a 6-lag VAR in order not to truncate the lag length and leave residual serial correlation in the model.

The VAR is unrestricted—that is, we impose no structural or exogeneity assumptions, even given the results of Section 3.2.4. This is critical because the VAR approach is not based on using the exogeneity tests to restrict the model and such restrictions may not be valid for the structural model. To understand which dynamic relationships may be part of the model, we analyze the model's impulse responses. These are computed from the estimated, unrestricted VAR. The estimated VAR is inverted and its moving average response (MAR) is presented in Figure 3.5, which replicates Figure 2 of Williams and Collins (1997). The error bands in the figure differ slightly from those reported in Williams and Collins because they are based on the likelihood-based bands described in Sims and Zha (1998). The error bands shown in Williams and Collins (1997) are based on a normality approximation, whereas those in Figure 3.5 are based on the actual 90% quantiles from a Monte Carlo simulation of the impulse responses. The results based on the quantiles are slightly larger—reflecting not only the parameter uncertainty but also uncertainty about the overall likelihood, shape, and skewness of the MARs.

The rows of the MAR graph show the response of a 1 standard deviation shock to the error in the equation for the column variable. This shock is then fed through the system of (inverted) VAR equations to produce the responses. The diagonal graphs show the responses of the variable to their own shocks, whereas the off-diagonal graphs show the responses of the variables to shocks in each other. The order in which these shocks enter the system is determined by the ordering of the variables in the decomposition of the shocks. Thus, shocks to ECTR enter first, followed by those to real GDP, then real investment, and finally those to the corporate PAC ratio.

The MAR plot supports Proposition 1, because a shock to ECTR leads to an increase in the ratio of corporate to business PACs. Proposition 2 shows that investment response to shocks in ECTR receives weak support; the general response is negative, but its 90% confidence interval includes zero. Finally, the MAR supports Proposition 3 because the 90% confidence bands for the response of ECTR to shocks in investment and income include zero. This means that ECTR does not respond to shocks or innovations in real income and real investment. Note also that the shock to the ratio of corporate PACs leads to an *increase* in ECTR, albeit one that does not differ from zero after eight quarters. An explanation offered by Williams and Collins (1997, p. 230) for this effect is that this increase in ECTR as a response to shocks in the corporate PAC ratio may be expectational. As the economy improves after shocks to the PAC ratio, this may lead to individuals and businesses to expect that tax rates will subsequently increase. In part, this reflects unmodeled expectations and dynamics, because we would not expect the real income and real investment to respond to the shocks in the corporate PAC ratio, which they do in Figure 3.5.

The MARs provide a clear, graphical summary of the complex dynamic system used to explain the four variables in this analysis. One major benefit of this method is that it allows the researcher to analyze how dynamic changes in a system of equations are related to one another. A second benefit is that the possible identification and exogeneity restrictions that would have been used in a structural equation representation of the system are

Shock To

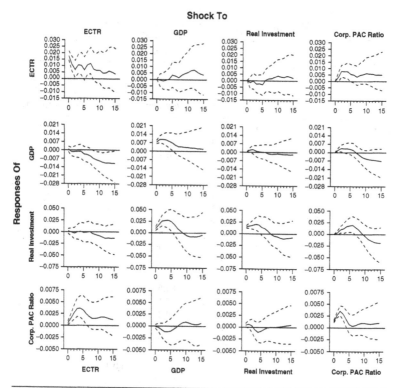

Figure 3.5 Moving Average Responses for Four-Variable VAR With 90% Error Bands, 1977–1994

NOTE: Responses are plotted on a 16-quarter horizon.

made explicit and can be tested. We see in the MAR how these exogeneity assumptions are or are not supported by the data. These results, combined with exogeneity or Granger causality analysis, provide a concise and clear way of addressing identification and dynamic specification in the same model.

3.2.6 Decomposition of the Forecast Error Variance

As a final demonstration of interpreting VAR models and their results, we present the decomposition of the forecast error variance analysis for the four-variable VAR discussed in Section 3.2.6. Table 3.12 presents the decomposition of the forecast error variance for the VAR.[12]

In Table 3.12, there are four blocks of forecast variance percentages—one for each variable in the system. In the first block, the forecast variance

TABLE 3.12
Decomposition of Error Variance for VAR

			Innovations In		
Forecast Error	k	ECTR	Investment	GDP	Ratio of Corporate PACs to Total Number of PACs
ECTR	1	100.0	0.0	0.0	0.0
	2	98.5	0.2	0.7	0.6
	3	93.2	0.4	0.7	5.8
	4	89.5	0.7	2.4	7.4
	6	83.5	5.0	2.7	8.8
	8	77.7	5.4	8.1	8.8
	10	74.2	5.7	11.7	8.5
	12	68.9	9.8	13.2	8.1
	16	63.4	14.7	13.2	8.8
GDP	1	0.0	99.9	0.0	0.0
	2	5.4	90.9	1.5	2.3
	3	8.4	85.5	3.2	3.0
	4	10.9	83.6	2.8	2.7
	6	17.7	77.1	2.5	2.8
	8	25.2	69.8	2.5	2.6
	10	32.3	63.0	2.2	2.4
	12	36.7	59.0	2.1	2.3
	16	40.1	54.1	3.6	2.2
Investment	1	0.0	39.6	60.3	0.0
	2	1.8	57.5	40.7	0.1
	3	4.6	66.1	28.4	0.9
	4	4.8	70.3	23.7	1.2
	6	6.7	73.4	17.6	2.4
	8	9.6	71.4	15.5	3.6
	10	13.9	68.0	14.6	3.5
	12	16.3	66.3	14.0	3.4
	16	17.0	65.5	14.0	3.4
Ratio of corporate PACs to total number of PACs	1	4.0	2.5	4.4	89.1
	2	8.4	2.6	2.5	86.5
	3	12.0	5.0	1.2	81.8
	4	17.6	7.9	2.7	71.9
	6	20.4	12.0	5.3	62.3
	8	18.8	18.5	5.2	57.5
	10	17.6	24.6	4.9	53.0
	12	19.4	25.4	4.8	50.4
	16	22.8	26.8	5.0	45.3

NOTE: Entries are percentage of forecast error resulting from innovations. Each entry represents the percentage of the forecast error k quarters ahead in the row variable that is due to innovations in that column variable.

for ECTR that is attributed to each of the four variables in the system is presented. Note that each row corresponding to a period in the forecast horizon sums to 100%. So the initial shock to ECTR attributes all the innovation in ECTR to ECTR itself, which is consistent with being the first variable in the ordering of the Cholesky decomposition of the responses. Then we trace through the system of equations for periods 2 through 16 to see whether the subsequent forecast variation in ECTR is due to innovations in ECTR or the other variables. By period 12, we see that about 30% of the total variation in ECTR is due to the innovations in the investment (9.8%), GDP (13.2%), and ratio of corporate PACs (8.1%) variables.

For the economic variables—GDP and investment—the innovations in investment explain most of the forecast variance. After 12 to 16 quarters, over half of the forecast variance in GDP is the result of innovations in investment and 40% are the result of innovations in ECTR. Thus, GDP responds primarily to these two variables. For the forecast errors in investment, we see that most of the errors are explained by innovations in investment itself—over 65% of the variance in investment is due to its own innovations after 12 periods.

There are several key conclusions to draw from these innovation accounting results. First, shocks to ECTR lead to a change in investment patterns. Innovations in ECTR account for about 17% of the forecast variance in investment after 16 quarters. Second, the impacts of innovations in ECTR on the ratio of corporate PACs to the total number of PACs is also large because about 23% of the variance in the PAC ratio can be attributed to innovations in ECTR. Note also that the ECTR does not have any appreciable response to the innovations in the other variables. Even after 16 quarters, nearly two thirds of the forecast error variance in ECTR is attributable to its own innovations. These results support Propositions 1 to 3.

3.2.7 A Further Robustness Check

One of the contributions of the Williams and Collins (1997) model and analysis is that effective corporate tax rates should be predictive of future economic events. This follows rather directly from their empirical analysis: Changes in ECTR are shown to affect corporate investment and real income (GDP). If ECTR and consequently tax policy predict investment and economic growth, then beliefs about taxes should aid in the prediction of *consumer expectations* about the direction of the economy. Williams and Collins note that the public's expectation of the direction of the economy—consumer expectations—should then be predicted by the VAR including the other variables.

One of the reasons this is an interesting check on the earlier analysis is that the PAC measures are available only for the period after 1976. A measure of consumer expectations, the index of consumer sentiment collected by the University of Michigan, is available from 1953. Williams and Collins (1997) evaluate the prediction and check the robustness of the results about the relationship of ECTR to investment and real income using a new four-variable VAR. This second VAR model uses ECTR, logged real income, logged real interest rates, and consumer expectations as variables. This analysis omits the PAC measure, so it does not allow us to analyze Proposition 1. The main focus here is analysis of Propositions 2 and 3 over the extended time period and the evaluation of the relationship between ECTR and consumer expectations. This also serves as a check on the stability of the relationships found earlier using a longer series of data.

Rather than present a full analysis or the Granger causality results (Williams & Collins, 1997, Table 5), here we present only the MAR. Figure 3.6 presents the MAR for this new VAR model with 6 lags.

The upper left 3×3 graphs in Figure 3.6 are nearly identical to those in Figure 3.5. This figure demonstrates the robustness of the earlier findings in several ways. First, the longer time horizon clarifies the impacts of ECTR on investment. The increase in ECTR leads to a decline in real investment, consistent with Proposition 2. The 1.5% shock to ECTR leads to a long-run response of about a 1% decline in real investment. Also, we see that the shocks to ECTR influence consumers' expectations: Each 1.5% shock in ECTR leads to a statistically significant 2.5-point decline in consumer sentiment after six quarters. Conversely, shocks to consumer sentiment do not lead to a response in ECTR, because the confidence interval for this response includes zero. These results are consistent with Granger causality results reported in Williams and Collins (1997).

The results of these VAR analyses support the three propositions of Williams and Collins (1997). The results are robust as well to changes in the sample, lag length, and measures of PAC or business political strength. Finally, the robustness check in this section supports the idea that effective corporate tax rates change in concert with economic aggregates and affect consumer expectations.

3.3 Conclusion

The two examples in this chapter are meant to illustrate the basic VAR approach. These methods can be applied to larger systems of variables (e.g., Leeper et al., 1996) and systems with more complex dynamics. Our

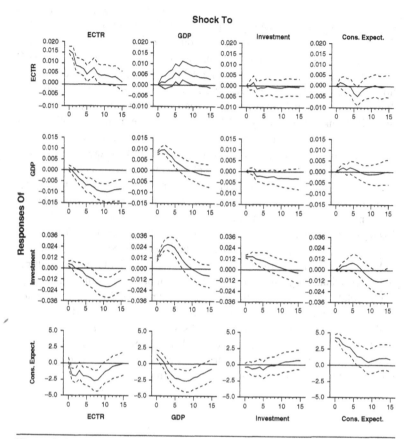

Figure 3.6 Moving Average Responses for Four-Variable VAR With 90% Error Bands, 1953–1994

NOTE: Responses are plotted on a 16-quarter horizon.

main goal here has been to illustrate how the VAR approach can be used to answer questions that may or may not have rigorous dynamic theories behind them.

In addition, it should be noted that the presentation here only scratches the surface of multiple time series modeling methods. VAR, SEQ, and ECM models are all regularly used for dynamic modeling, forecasting, and policy analysis. Forecasting and policy analysis are the subject of the seminal paper by Doan, Litterman, and Sims (1984), with a more recent Bayesian treatment by Waggoner and Zha (1999) and less technical treatments in Brandt and Freeman (2006) and Zha (1998).

There also exist many extensions to these methods that allow for structural modeling using VARs (structural VARs or SVARs). In SVARs, theory is used to identify the contemporaneous innovations in a VAR model in the same way as SEQ models. This has led to SVAR models that use SEQ modeling techniques to construct a model of the contemporaneous innovations. These models have been used to identify aggregate supply and demand relationships (Blanchard & Quah, 1989), model the macroeconomy (Bernanke, 1986; Sims, 1981, 1986a), analyze monetary policy rules (Leeper et al., 1996), and describe the relationship between presidential approval and the macroeconomy (Williams, 1990). Waggoner and Zha (2003) is a more recent theoretical development of these models.

In addition, there is the large literature on error correction models and vector error correction models. Lutkepohl (2004) offers an accessible review of these models and their relationships to the VAR models discussed here. Exemplar social science applications (non-economics) include Ostrom and Smith (1993), Erikson et al. (2002), and Clarke, Ho, and Stewart (2000).

Finally, even if one uses a VAR (or Bayesian VAR) to model multiple time series data, the goal is the characterization of the dynamics and endogenous relationships among the time series. This book has demonstrated some of the main ways that one can use to characterize the dynamics of multiple time series and related models such as simultaneous equation, error correction, and structural models. Whichever approach is used, the goals are the same: showing whether various variables are endogenously related, with what dynamics, and over what time periods.

APPENDIX: SOFTWARE FOR MULTIPLE
TIME SERIES MODELS

Working with multiple time series data typically can be done with standard regression software. That is, if all one wishes to do is estimate ARIMA or simultaneous equation (SEQ) models, many standard statistical software programs contain the relevant procedures for the estimation and hypothesis tests. However, as most analysts of time series data quickly discover, standard packages are woefully inadequate at recognizing the time series properties of data. With this in mind, we present some criteria that one should use in selecting software for analyzing multiple time series data and a brief list of some software that is available for analyzing multiple time series data.

The first thing to recognize about any time series data project is that we want to do different things with time series data (Brandt, 2002). First, one will want to organize and subset the time series data for relevant time samples. Often, one collects data over differing time periods and then has to splice these data together. Although this can commonly be done in most relational database or statistical software programs, subsetting the data with respect to date is important. This requires that the software has a way to recognize that the variables in the data set are themselves a time series and to know how to subset the data. Good time series software will make this easy.

Second, time series software should be able to recognize, construct, and manipulate sequences of dates and data that have time units. This includes functions for things such as generating leads and lagged values of variables. Another important piece of functionality (especially in recent years) has been provisions for converting two-digit year dates to four-digit year dates (in the wake of moving from the 20th to the 21st century). Being able to work with dates and time units will typically be accompanied by specialized graphical functions for plotting time series (or even multiple time series) on the same graph with a common time scale.

Third, for SEQ and VAR models, hypothesis tests and fit measures are different than in standard regression models. Tests and diagnostics for autocorrelation, Granger causality, lag length selection, unit roots, and error correction models are commonly present in good multiple time series software.

Fourth, more complex and model-specific computations are available only in time series software. It is generally possible to use standard linear regression or multivariate linear regression software to estimate a VAR or multistage simultaneous equation model. But the quantities of interest for these models—dynamic simulations, impact multipliers, impulse responses, innovation accounting, Monte Carlo sampling for error band construction, and so on—are typically available only in highly specialized time series software. Thus, rather than spend time trying to cajole a program into doing something, one may want to investigate which software program or programming language provides the functionality needed for the analysis one wishes to perform.

Finally, our own experiences in teaching, analyzing, and developing multiple time series models indicate that using good time series software makes the analysis faster, more efficient, and less prone to error. Thus, any choice of software for multiple time series modeling should be one that is familiar and easy to learn and facilitates the desired modeling and analysis.

Our goal is not to suggest "one" program or statistical package for (multiple) time series analysis. Rather, it is to point out that analysis may proceed with any package with which someone is familiar. But note that there may be limitations in the selection of the software and that these limitations should not be allowed to define the choice of how analysis is conducted. If one wishes to test hypotheses about the impacts of innovations in VAR models, then it is likely that they should use software that can estimate impulse responses, innovation accounting, and the variability of these quantities.

In our experience, a large number of statistical analysis software can be used for analyzing the multiple time series models we discuss. As specialists in time series analysis, our evaluation of the software based on the criteria outlined above leads us to segment the currently available products into three major groups. The first includes general programs that can recognize that the data are multiple time series and include some of the basic tools for analyzing VAR and SEQ models. The second includes more specialized programs that are specifically designed to analyze time series data (sometimes almost exclusively). These programs include specialized hypothesis testing and model-fit measures for multiple time series models. Forecasting is readily available in these programs. The final group of programs includes high-level programming languages that can best be used to program highly specialized multiple time series models because they contain powerful matrix algebra and time series functions. Here, we provide a brief overview that can be used as a starting basis for determining which type of software program may meet a user's needs.

Software in the first group include programs like SPSS, SAS, and Stata. These are general programs commonly used for statistical analysis in the social sciences and can be accessed with a graphical user interface. All are capable of some time series analysis. SAS and Stata include specific procedures for estimating VAR models, impulse responses, innovation accounting, hypothesis testing for lag length, and Granger causality tests. In Stata, the var and var intro functions can be used to estimate and find commands for interpreting VAR models. In SAS, the VARMAX procedure allows one to estimate VARs, Bayesian VARs, and VECMs. SPSS is more limited in this regard, because its ability to estimate simultaneous equation models (via AMOS) is not focused on time series models.

There are several statistical software programs designed specifically for multiple time series analysis. These include programs that have graphical interfaces, require writing code, or allow both modes of interaction. Estima's RATS (Regression Analysis of Time Series) program and CATS (Cointegration Analysis of Time Series) procedures present a comprehensive programming environment for analyzing univariate and multivariate time series models (ARIMA, SEQ, VAR, VECM, etc.). It is probably the most comprehensive program for the analysis of VAR models and is written by Thomas Doan, one of the authors of the paper by Doan et al. (1984). RATS is quite powerful, but requires writing code rather than using a graphical interface. RATS can estimate all varieties of VAR models, impulse responses, innovation accounting, Bayesian VAR models, SVAR models, and much more. It can also be extended to analyze customized models via a native matrix programming language.

The program JMulti is another program for multiple time series models. It is described in the book by Lutkepohl and Kratzig (2004) and can be used for VAR and VECM models. It includes extensive hypothesis testing methods and model interpretation via impulse responses and innovation accounting. Eviews is a time series analysis program that includes both strong graphical user interface and programming language. It is capable of testing most univariate and multivariate time series models described here. Finally, Ox and Oxmetrics constitute a series of programs that can be used to estimate VAR and VECM models (among others). Ox is a comprehensive environment for time series models that can be accessed either by graphical interfaces or by running user written code. Additional packages are available for the Markov-switching models described in Krolzig (1997) and other time series models.

The final group of programs consists of high-level statistical analysis and programming languages, including Aptech's Gauss program for matrix algebra and statistical analysis, Insightful's Splus program, which includes a graphical interface as well as the power S language for statistical and graphical data analysis, the open-source GNU clone of Splus known as R, and the matrix algebra and data analysis program Matlab. These are powerful programs that also include "libraries," "packages," and "toolboxes" that include procedures for analyzing multiple time series with VAR, VECM, and other methods. As high-level programming languages, these programs typically require writing code to estimate the models. But this "expense" (i.e., not being able to do everything with a graphical interface) brings with it additional power. These programs, however, are very general and powerful and therefore often do not include some of the more detailed time series functions and features described in the programs specifically designed for time series analysis.

Details and additional information about these programs can be easily found on the Internet. All the programs described above can be run on Microsoft Windows operating systems and are also available for Apple OS X and Unix/Linux operating systems.

NOTES

Chapter 1

1. A notable exception is the supply and demand model in economics.

2. The basic requirement is known as an order condition: The number of exogenous or predetermined variables not included in an equation must be greater than the number of total endogenous variables minus one.

3. This parallels standard definitions of identification and structural models (Judge, Griffiths, Hill, Lutkepohl, & Lee, 1985, pp. 573–574).

4. This argument is made cogently by Freeman et al. (1989).

5. The main point here, familiar to users of linear regression or simultaneous equation models, is that the matrix of the right-hand side variables must be "full rank" or meet a rank and order condition. Although in practice there are rules to ensure identification, even models that satisfy these rank and order rules may be unidentified or observationally equivalent to other representations in the sense that their parameters are not a unique representation of the quantities of interest.

6. More generally, a series integrated to order d, denoted as $I(d)$, requires d differencing operations to be made stationary. Series of similar orders d can be modeled together.

7. One can also include moving average components and construct a vector autoregressive moving average model (see Lutkepohl, 2005).

Chapter 2

1. The Kronecker product of an $(m \times n)$ matrix A and a $(p \times q)$ matrix B is defined as the following $(mp) \times (nq)$ matrix:

$$A \otimes B = \begin{bmatrix} a_{11}B & a_{12}B & \cdots & a_{1n}B \\ a_{21}B & a_{22}B & \cdots & a_{2n}B \\ \vdots & \vdots & \cdots & \vdots \\ a_{m1}B & a_{m2}B & \cdots & a_{mn}B \end{bmatrix}.$$

2. Details of these derivations and the maximum-likelihood solution can be found in Hamilton (1994, chap. 11).

3. This is actually an important point: Deseasonalized data are typically filtered using an ARIMA or $X11$ procedure that can actually amplify some of the seasonal serial correlations. Capturing enough lags in the VAR helps ensure that these patterns are modeled correctly.

4. Using too many lags, which are highly correlated, leads to an $(X'X)$ that may be singular and noninvertible, thus making it impossible to compute the OLS estimates.

5. Note that in the computation of the likelihood ratio test, unlike in the likelihood functions, we do NOT use the inverses of the error covariance Σ. This is because of the following simplifications (using the properties of determinants and logarithms):

$$
\begin{aligned}
2\big(L\big(\hat{\Sigma}, B, p_1\big) - L\big(\hat{\Sigma}, B, p_0\big)\big) &= 2\left[\frac{T}{2}\log|\widehat{\Sigma_1^{-1}}| \frac{T}{2}\log|\widehat{\Sigma_0^{-1}}|\right] \\
&= T\left(\log\left(\frac{1}{|\widehat{\Sigma_1}|}\right) - \log\left(\frac{1}{|\widehat{\Sigma_0}|}\right)\right) \\
&= -T\left(-\log|\widehat{\Sigma_1}| + \log|\widehat{\Sigma_0}|\right) \\
&= T\left(\log|\widehat{\Sigma_0}| - \log|\widehat{\Sigma_1}|\right).
\end{aligned}
$$

6. Typical adjustments for multiple tests such as this are either Bonferroni-adjusted P-values or Šidák-adjusted P-values. The Bonferroni-adjusted values for the kth test in a sequence are $p_i^b = min(1, k p_i)$, where k is the number of hypotheses tested and p_i are the unadjusted P-values (say, 0.05). The Šidák-adjusted P-values for the kth test in a sequence are $p_i^s = 1 - (1 - p_i)^k$.

7. This can be seen by comparing $\text{AIC}(p + 1)$ with $\text{AIC}(p)$ under the assumption that there is no change in the $\log|\hat{\Sigma}|$. The difference of these two will be a factor of m^2. So unless the $\log|\hat{\Sigma}|$ across the model with $p + 1$ and p lags changes by more than the penalty of m^2 more parameters, one will choose the more parsimonious model.

8. The trace of a square matrix is the sum of its diagonal elements.

9. For an alternative method, see Sims (1972) and the discussion in Freeman (1983).

10. This section is adapted from a paper on the topic of assessing uncertainty about dynamics in VAR models by Brandt and Freeman (2006). Brandt would like to thank Freeman for allowing this adaptation of the material they developed together.

11. For an introduction to Monte Carlo sampling methods and their applications, see Mooney (1997).

12. Error correction models are rarely applied for stationary time series data, because without a stochastic trend, there is little risk of estimating a spurious regression. However, many key economic and social variables have unit root or stochastic trend properties and may need to consider the possibility of modeling these as unit root or near unit root processes (DeBoef & Granato, 1997).

13. Note that this would also be the same representation for a VECM system of more than two variables, because y_t could be a matrix of more than two variables.

14. See DeBoef and Granato (1997) for an alternative interpretation.

Chapter 3

1. The data and RATS code used to generate these examples are available from the lead author's Web page and the publisher's Web site.

2. Erikson et al. (2002) present a set of error correction and dynamic simulations that analyze these variables as well.

3. Alternatively, if one believes that macropartisanship—a nonnegative variable—could not *theoretically* have a unit root, one could just eschew using unit root testing on this ground alone.

4. Strictly holding to the AIC value that will best fit the data (the smallest value of the AIC and p) tends to select models that have too many lags (Lutkepohl, 2004).

5. Rerunning the model and results in subsequent sections with 2 lags produces nearly identical results.

6. For longer time horizons, one might just present the decomposition of the forecast error variance results for key dates as we do in the second example.

7. Changing the ordering alters the normalization of the Cholesky decomposition and the order in which the equations are shocked in the computation of the moving average response.

8. For an analysis of similar data that focus on the unit roots and the possible cointegrating relationships, see Inclan, Quinn, and Shapiro (2001). For an alternative VAR method for addressing these data, see Freeman et al. (1998).

9. Using the Bonferroni- and Šidák-adjusted P-values described in Chapter 2, a case can be made for using an 8-lag model. Replicating this model and subsequently using 8 lags produces similar results; thus, we choose to present the more parsimonious model.

10. Bivariate VARs produce asymptotically correct hypothesis test results even in the presence of unit roots. If unit roots are present, the robustness of the results can be assessed by using a VAR in first differences.

11. Note that these results differ slightly from Table 3 of Williams and Collins (1997). There they incorrectly report some of the 8-lag results as 4-lag results. Their original conclusions still hold, however.

12. These results are analogous but not identical to those in Table 4 of Williams and Collins (1997). Their decomposition of the forecast error variance is based on a time-varying Bayesian VAR, whereas the results reported here are based on an unrestricted, non-time-varying VAR without a Bayesian prior.

REFERENCES

Banerjee, A., Dolado, J. J., Galbraith, J. W., & Hendry, D. F. (1993). *Co-integration, error correction and the econometric analysis of non-stationary data.* Oxford, UK: Oxford University Press.

Bernanke, B. (1986). Alternative explanations of the money-income correlation. In *Carnegie-Rochester conference series on public policy.* Amsterdam: North-Holland.

Blanchard, O., & Quah, D. (1989). The dynamic effects of aggregate demand and supply disturbances. *American Economic Review, 79,* 655–673.

Box, G. E. P., & Jenkins, G. M. (1970). *Time series analysis, forecasting and control.* San Francisco: Holden Day.

Box-Steffensmeier, J. M., & Smith, R. M. (1996). The dynamics of aggregate partisanship. *American Political Science Review, 90*(3), 567–580.

Brandt, P. T. (2002). Using the right tools for time series data analysis. *Political Methodologist, 10*(2), 22–26.

Brandt, P. T., & Freeman, J. R. (2006). Advances in Bayesian time series modeling and the study of politics: Theory testing, forecasting, and policy analysis. *Political Analysis, 14*(1), 1–36.

Clarke, H. D., Ho, K., & Stewart, M. C. (2000). Major's lesser (not minor) effects: Prime ministerial approval and governing party support in Britain since 1979. *Electoral Studies, 19,* 255–273.

Cooley, T. F., & LeRoy, S. F. (1985). Atheoretical macroeconomics: A critique. *Journal of Monetary Economics, 16,* 283–308.

Cromwell, J. B., Labys, W. C., Hannan, M. J., & Terraza, M. (1994). *Multivariate tests for time series models.* Thousand Oaks, CA: Sage.

DeBoef, S., & Granato, J. (1997). Near integrated data and the analysis of political relationships. *American Journal of Political Science, 41*(2), 619–640.

Doan, T., Litterman, R., & Sims, C. (1984). Forecasting and conditional projection using realistic prior distributions. *Econometric Reviews, 3,* 1–100.

Dolado, J. J., & Lutkepohl, H. (1996). Making Wald tests work for cointegrated VAR systems. *Econometric Reviews, 15*(4), 369–386.

Engle, R. F., & Granger, C. W. J. (1987). Co-integration and error correction: Representation, estimation and testing. *Econometrica, 55,* 251–276.

Erikson, R. S., MacKuen, M. B., & Stimson, J. A. (2002). *The macropolity.* New York: Cambridge University Press.

Freeman, J. R. (1983). Granger causality and the time series analysis of political relationships. *American Journal of Political Science, 27*(3), 327–358.

Freeman, J. R., Williams, J. T., Houser, D., & Kellstedt, P. (1998). Long memoried processes, unit roots and causal inference in political science. *American Journal of Political Science, 42*(4), 1289–1327.

Freeman, J. R., Williams, J. T., & Lin, T.-M. (1989). Vector autoregression and the study of politics. *American Journal of Political Science, 33,* 842–877.

Granger, C. W. J. (1969). Investigating causal relations by econometric models and crossspectral methods. *Econometrica, 37,* 424–438.

Granger, C. W. J., & Newbold, P. (1974). Spurious regressions in econometrics. *Journal of Econometrics, 2,* 111–120.

Granger, C. W. J., & Newbold, P. (1986). *Forecasting economic time series* (2nd ed.). San Diego, CA: Academic Press.

Hamilton, J. D. (1994). *Time series analysis*. Princeton, NJ: Princeton University Press.

Harvey, A. C. (1990). *The econometric analysis of time series* (3rd ed.). Cambridge, MA: MIT Press.

Hatanaka, M. (1975). On the global identification of the dynamic simultaneous equations model with stationary disturbances. *International Economic Review, 16*(3), 545–554.

Inclan, C., Quinn, D. P., & Shapiro, R. Y. (2001). Origins and consequences of changes in US corporate taxation, 1981–1998. *American Journal of Political Science, 45*(1), 179–201.

Johansen, S. (1995). *Likelihood-based inference in cointegrated vector autoregressive models*. Oxford, UK: Oxford University Press.

Judge, G., Griffiths, W. E., Hill, R. C., Lutkepohl, H., & Lee, T.-C. (1985). *Theory and practice of econometrics* (2nd ed.). New York: Wiley.

Kilian, L. (1998). Small-sample confidence intervals for impulse response functions. *Review of Economics and Statistics, 80*, 186–201.

Krolzig, H,-M. (1997). *Markov-switching vector autoregressions: Modelling, statistical inference, and application to business cycle analysis*. Berlin: Springer.

Kwiatkowski, D., Phillips, P. C. B., Schmidt, P., & Shin, Y. (1992). Testing the null hypothesis of stationarity against the alternative of a unit root. *Journal of Econometrics, 54*, 159–178.

Leeper, E. M., Sims, C. A., & Zha, T. (1996). What does monetary policy do? *Brookings Papers on Economic Activity, 2*, 1–63.

Litterman, R. B., & Weiss, L. (1985). Money, real interest rates, and output: A reinterpretation of postwar U.S. data. *Econometrica, 53*(1), 129–156.

Lutkepohl, H. (1985). Comparison of criteria for estimating the order of a vector. *Journal of Time Series Analysis, 6*, 35–52. (Correction: 1987, 8, 373)

Lutkepohl, H. (1990). Asymptotic distributions of impulse response functions and forecast error variance decompositions in vector autoregressive models. *Review of Economics and Statistics, 72*, 53–78.

Lutkepohl, H. (2004). Vector autoregressive and vector error correction models. In H. Lutkepohl & M. Kratzig (Eds.), *Applied time series econometrics* (pp. 86–158). Cambridge, UK: Cambridge University Press.

Lutkepohl, H. (2005). *New introduction to multiple time series analysis*. Berlin: Springer.

Lutkepohl, H., & Kratzig, M. (Eds.). (2004). *Applied time series econometrics*. Cambridge, UK: Cambridge University Press.

Lutkepohl, H., & Reimers, H.-E. (1992a). Granger-causality in cointegrated VAR processes: The case of term structure. *Economics Letters, 40*, 263–268.

Lutkepohl, H., & Reimers, H.-E. (1992b). Impulse response analysis of cointegrated systems. *Journal of Economic Dynamics and Control, 16*, 53–78.

Mills, T. (1991). *Time series techniques for economists*. Cambridge, UK: Cambridge University Press.

Mittnik, S., & Zadrozny, P. A. (1993). Asymptotic distributions of impulse responses, step responses, and variance decompositions of estimated linear dynamic models. *Econometrica, 20*, 832–854.

Mooney, C. Z. (1997). *Monte Carlo simulation*. Thousand Oaks, CA: Sage.

Ostrom, C. (1990). *Time series analysis: Regression techniques*. Thousand Oaks, CA: Sage.

Ostrom, C., & Smith, R. (1993). Error correction, attitude persistence, and executive rewards and punishments: A behavioral theory of presidential approval. *Political Analysis, 3*, 127–184.

Pagan, A. (1987). Three econometric methodologies: A critical appraisal. *Journal of Economic Surveys, 1*(1), 3–24.

Philips, C. B. (1986). Understanding spurious regressions in econometrics. *Journal of Econometrics, 33*, 311–340.

Reinsel, G. C. (1993). *Elements of multivariate time series analysis*. New York: Springer–Verlag.

Runkle, D. E. (1987). Vector autoregressions and reality. *Journal of Business and Economic Statistics, 5*, 437–442.

Saikkonen, P., & Lutkepohl, H. (1999). Local power of likelihood ratio tests for the cointegrating rank of VAR processes. *Econometric Theory, 15*, 50–78.

Saikkonen, P., & Lutkepohl, H. (2000a). Testing for cointegrating rank of a VAR process with an intercept. *Econometric Theory, 16*, 373–406.

Saikkonen, P., & Lutkepohl, H. (2000b). Trend adjustment prior to testing for the cointegrating rank of a vector autoregressive process. *Journal of Time Series Analysis, 21*, 435–456.

Sims, C. A. (1972). Money, income, and causality. *American Economic Review, 62*, 540–552.

Sims, C. A. (1980). Macroeconomics and reality. *Econometrica, 48*(1), 1–48.

Sims, C. A. (1981). An autoregressive index model for the U.S., 1948–1975. In J. Kmenta & J. B. Ramsey (Eds.), *Large-scale macro-econometric models* (pp. 283–327). Amsterdam: North-Holland.

Sims, C. A. (1986a). Are forecasting models usable for policy analysis? *Quarterly Review, Federal Reserve Bank of Minneapolis, 10*, 2–16.

Sims, C. A. (1986b). Specification, estimation, and analysis of macroeconomic models. *Journal of Money, Credit and Banking, 18*(1), 121–126.

Sims, C. A. (1987). Comment [on Runkle]. *Journal of Business and Economic Statistics, 5*(4), 443–449.

Sims, C. A. (1988). Uncertainty across models. In *Papers and proceedings of the one-hundredth annual meeting of the American Economic Association* (pp. 163–167). Nashville, TN: American Economic Association American Economic Review.

Sims, C. A. (1996). Macroeconomics and methodology. *Journal of Economic Perspectives, 10*, 105–120.

Sims, C. A., Stock, J. H., & Watson, M. W. (1990). Inference in linear time series models with some unit roots. *Econometrica, 58*(1), 113–144.

Sims, C. A., & Uhlig, H. (1991). Understanding unit rooters: A helicopter tour. *Econometrica, 59*(6), 1591–1599.

Sims, C. A., & Zha, T. (1998). Bayesian methods for dynamic multivariate models. *International Economic Review, 39*(4), 949–968.

Sims, C. A., & Zha, T. (1999). Error bands for impulse responses. *Econometrica, 67*(5), 1113–1156.

Stimson, J. (1999). *Public opinion in America: Moods, cycles, and swings* (2nd ed.). Boulder, CO: Westview.

Toda, H. Y., & Phillips, P. C. B. (1993). Vector autoregressions and causality. *Econometrica, 61*(6), 1367–1393.

Toda, H. Y., & Yamamoto, T. (1995). Statistical inference in vector autoregressions with possibly integrated processes. *Journal of Econometrics, 66*, 225–250.

Waggoner, D. F., & Zha, T. (1999). Conditional forecasts in dynamic multivariate models. *Review of Economic and Statistics, 81*(4), 639–651.

Waggoner, D. F., & Zha, T. A. (2003). A Gibbs sampler for structural vector autoregressions. *Journal of Economic Dynamics & Control, 28*, 349–366.

Williams, J. T. (1990). The political manipulation of macroeconomic policy. *American Political Science Review, 84*(3), 767–795.

Williams, J. T. (1993). What goes around comes around: Unit root tests and cointegration. *Political Analysis, 4*, 229–235.

Williams, J. T., & Collins, B. K. (1997). The political economy of corporate taxation. *American Journal of Political Science, 41*(1), 208–244.

Williams, J. T., & McGinnis, M. D. (1988). Sophisticated reaction in the U.S.-Soviet arms race: Evidence of rational expectations. *American Journal of Political Science, 32*(4), 968–995.

Wold, H. (1954). *A study in the analysis of stationary time series* (2nd ed.). Uppsala, Sweden: Almqvist and Wiksell.

Zapata, H. O., & Rambaldi, A. N. (1997). Monte Carlo evidence on cointegration and causation. *Oxford Bulletin of Economics and Statistics, 59*(2), 285–298.

Zellner, A. (1971). *An introduction to Bayesian inference in econometrics.* New York: Wiley Interscience.

Zellner, A., & Palm, F. C. (Eds.). (2004). *The structural econometric time series analysis approach.* Cambridge, UK: Cambridge University Press.

Zha, T. (1998). A dynamic multivariate model for the use of formulating policy. *Economic Review (Federal Reserve Bank of Atlanta), 83*(1), 16–29.

INDEX

ABOUT THE AUTHORS

Patrick T. Brandt is Assistant Professor of Political Science in the School of Economic, Political, and Policy Sciences at the University of Texas at Dallas. He has published in the *American Journal of Political Science* and *Political Analysis*. He teaches courses in social science research methods and social science statistics. His current research focuses on the development and application of time series models to the study of political institutions, political economy, and international relations. Before joining the faculty at the University of Texas at Dallas, he held positions at the University of North Texas and Indiana University, and was a fellow at the Harvard-MIT Data Center. He received an AB (1990) in Government from the College of William and Mary, an MS (1997) in Mathematical Methods in the Social Sciences from Northwestern University, and a PhD (2001) in Political Science from Indiana University.

John T. Williams was Professor and Chair of the Department of Political Science at the University of California, Riverside. He taught time series analysis at the Inter-University Consortium for Political and Social Research Summer Training Program for over 10 years. His work uses statistical methods in the study of political economy and public policy. He coauthored two books: *Compound Dilemmas: Democracy, Collective Action, and Superpower Rivalry* (2001) and *Public Policy Analysis: A Political Economy Approach* (2000). He published over 20 journal articles and book chapters on a wide range of topics, ranging from macroeconomic policy to defense spending to forest resource management. He was a leader in the application of new methods of statistical analysis to political science, especially the use of vector autoregression, Bayesian, and event count time series models. Before moving to Riverside in 2001, he held academic positions at the University of Illinois, Chicago (1985–1990), and Indiana University, Bloomington (1990–2001). He received a BA (1979) and an MA (1981) from North Texas State University and a PhD (1987) from the University of Minnesota.

Printed in the United States
By Bookmasters